BIOHYDROGEN

Developments and Prospects

BIOHYDROGEN

Developments and Prospects

Edited by
Dr. Sonil Nanda
Dr. Prakash K. Sarangi

First edition published 2022

Apple Academic Press Inc.
1265 Goldenrod Circle, NE,
Palm Bay, FL 32905 USA
4164 Lakeshore Road, Burlington,
ON, L7L 1A4 Canada

CRC Press
6000 Broken Sound Parkway NW,
Suite 300, Boca Raton, FL 33487-2742 USA
2 Park Square, Milton Park,
Abingdon, Oxon, OX14 4RN UK

© 2022 Apple Academic Press, Inc.

Apple Academic Press exclusively co-publishes with CRC Press, an imprint of Taylor & Francis Group, LLC

Reasonable efforts have been made to publish reliable data and information, but the authors, editors, and publisher cannot assume responsibility for the validity of all materials or the consequences of their use. The authors, editors, and publishers have attempted to trace the copyright holders of all material reproduced in this publication and apologize to copyright holders if permission to publish in this form has not been obtained. If any copyright material has not been acknowledged, please write and let us know so we may rectify in any future reprint.

Except as permitted under U.S. Copyright Law, no part of this book may be reprinted, reproduced, transmitted, or utilized in any form by any electronic, mechanical, or other means, now known or hereafter invented, including photocopying, microfilming, and recording, or in any information storage or retrieval system, without written permission from the publishers.

For permission to photocopy or use material electronically from this work, access www.copyright.com or contact the Copyright Clearance Center, Inc. (CCC), 222 Rosewood Drive, Danvers, MA 01923, 978-750-8400. For works that are not available on CCC please contact mpkbookspermissions@tandf.co.uk

Trademark notice: Product or corporate names may be trademarks or registered trademarks and are used only for identification and explanation without intent to infringe.

Library and Archives Canada Cataloguing in Publication

Title: Biohydrogen : developments and prospects / edited by Dr. Sonil Nanda, Dr. Prakash K. Sarangi.

Names: Nanda, Sonil, editor. | Sarangi, Prakash Kumar, editor.

Description: First edition. | Includes bibliographical references and index.

Identifiers: Canadiana (print) 20210389729 | Canadiana (ebook) 2021038977X | ISBN 9781774639801 (hardcover) | ISBN 9781774639818 (softcover) | ISBN 9781003277156 (ebook)

Subjects: LCSH: Hydrogen—Biotechnology. | LCSH: Hydrogen as fuel. | LCSH: Biomass energy.

Classification: LCC TP248.65.H9 B56 2022 | DDC 665.8/1—dc23

Library of Congress Cataloging-in-Publication Data

CIP data on file with US Library of Congress

ISBN: 978-1-77463-980-1 (hbk)
ISBN: 978-1-77463-981-8 (pbk)
ISBN: 978-1-00327-715-6 (ebk)

About the Editors

Sonil Nanda, PhD
Research Associate, Department of Chemical and Biological Engineering, University of Saskatchewan, Saskatoon, Saskatchewan, Canada

Dr. Sonil Nanda is a Research Associate in the Department of Chemical and Biological Engineering at the University of Saskatchewan, Saskatoon, Saskatchewan, Canada. He has published over 120 peer-reviewed journal articles, 70 book chapters and has presented at many international conferences. His research areas are related to the production of advanced biofuels and biochemical through thermochemical and biochemical conversion technologies such as gasification, pyrolysis, carbonization, torrefaction, and fermentation. He has gained expertise in hydrothermal gasification of various organic wastes and biomass, including agricultural and forestry residues, industrial effluents, municipal solid wastes, cattle manure, sewage sludge, food wastes, waste tires, and petroleum residues to produce hydrogen fuel. His similar interests are also in the generation of hydrothermal flames for the treatment of hazardous wastes, agronomic applications of biochar, phytoremediation of heavy metal contaminated soils, as well as carbon capture and sequestration.

Dr. Nanda is the editor of books entitled *New Dimensions in Production and Utilization of Hydrogen* (Elsevier), *Recent Advancements in Biofuels and Bioenergy Utilization* (Springer Nature), *Biorefinery of Alternative Resources: Targeting Green Fuels and Platform Chemicals* (Springer Nature), *Fuel Processing and Energy Utilization* (CRC Press), *Bioprocessing of Biofuels* (CRC Press) and *Biotechnology for Sustainable Energy and Products* (I.K. International Publishing House Pvt. Ltd.). Dr. Nanda is a Fellow Member of the *Society for Applied Biotechnology* in India and a Life Member of the *Indian Institute of Chemical Engineers, Association of Microbiologists of India, Indian Science Congress Association*, and the *Biotech Research Society of India*. He is also an active member of several chemical engineering societies across North America, such as the *American Institute of Chemical Engineers*, the *Chemical Institute of Canada*, the *Combustion Institute-Canadian Section*, and *Engineers Without Borders Canada*. Dr.

Nanda is Assistant Subject Editor of the *International Journal of Hydrogen Energy* (Elsevier) as well as an Associate Editor of *Environmental Chemistry Letters* (Springer Nature) and *Applied Nanoscience* (Springer Nature). He has also edited several special issues in renowned journals such as the *International Journal of Hydrogen Energy* (Elsevier), *Chemical Engineering Science* (Elsevier), *Biomass Conversion and Biorefinery* (Springer Nature), *Waste, and Biomass Valorization* (Springer Nature), *Topics in Catalysis* (Springer Nature), *SN Applied Sciences* (Springer Nature), and *Chemical Engineering and Technology* (Wiley).

Dr. Nanda received his PhD degree in Biology from York University, Canada; MSc degree in Applied Microbiology from Vellore Institute of Technology (VIT University), India; and BSc degree in Microbiology from Orissa University of Agriculture and Technology, India. He has worked as a Postdoctoral Research Fellow at York University, the University of Western Ontario and the University of Saskatchewan in Canada.

Prakash K. Sarangi, PhD
Scientist, Central Agricultural University, Imphal, Manipur, India

Dr. Prakash K. Sarangi is a Scientist with a specialization in Food Microbiology at the Central Agricultural University in Imphal, Manipur, India. His current research is focused on bioprocess engineering, renewable energy, biofuels, biochemicals, biomaterials, fermentation technology, and postharvest engineering and technology. He has more than 10 years of teaching and research experience in biochemical engineering, microbial biotechnology, downstream processing, food microbiology, and molecular biology.

Dr. Sarangi has served as a reviewer for many international journals and has authored more than 70 peer-reviewed research articles and 50 book chapters. Dr. Sarangi has edited the following books entitled *Recent Advancements in Biofuels and Bioenergy Utilization* (Springer Nature), *Biorefinery of Alternative Resources: Targeting Green Fuels and Platform Chemicals* (Springer Nature), *Fuel Processing and Energy Utilization* (CRC Press), *Bioprocessing of Biofuels* (CRC Press), and *Biotechnology for Sustainable Energy and Products* (I.K. International Publishing House Pvt. Ltd.). Dr. Sarangi serves as an academic editor for PLOS One journal. He is associated with many scientific societies as a fellow member (Society for Applied Biotechnology) and Life Member (Biotech Research Society of India; Society for Biotechnologists of India; Association of Microbiologists of India; Orissa Botanical Society; Medicinal and Aromatic Plants Association of India; Indian Science Congress Association; Forum of Scientists, Engineers, and Technologists; International Association of Academicians and Researchers; Hong Kong Chemical, Biological, and Environmental Engineering Society; International Association of Engineers; and Science and Engineering Institute).

Dr. Sarangi received his PhD degree in Microbial Biotechnology from the Department of Botany at Ravenshaw University, Cuttack, India; MTech degree in Applied Botany from the Indian Institute of Technology Kharagpur, India; and MSc degree in Botany from Ravenshaw University, Cuttack, India.

Contents

Contributors .... *xi*
Abbreviations .... *xiii*
Preface .... *xvii*

1. **Recent Advancements in Biohydrogen Production: Thermochemical and Biological Conversion Routes 1**
   Meenakshi Rajput, Amandeep Brar, Vivekanand Vivekanand, and Nidhi Pareek

2. **Production of Hydrogen through Gasification Technology 33**
   Paresh H. Rana

3. **Hydrogen Production by Catalytic Reforming Process 77**
   Chandramani Rai and Prabu Vairakannu

4. **Co-Conversion of Plastic Wastes and Biomass into Biohydrogen 109**
   Krushna Prasad Shadangi

5. **An Overview of Fermentative Hydrogen Production Technologies 127**
   Prakash K. Sarangi, Sonil Nanda, Ajay K. Dalai, and Janusz A. Kozinski

6. **Genetic Engineering and Fabrication of Microbial Cell System for Biohydrogen Production 141**
   Sushma Chauhan, Balasubramanian Velramar, Rakesh Kumar Soni, Mohit Mishra, Vargobi Mukherjee, Tanushree Baldeo Madavi, and Sudheer D.V.N. Pamidimarri

7. **Hydrogen Production through Microbial Electrolysis 175**
   Latika Bhatia, Prakash K. Sarangi, and Sonil Nanda

8. **Confluence of Nanocatalysts and Bioenergy: An Overview of Microbial Electrochemical Systems and Biohydrogen Production 189**
   Piyush Parkhey, Kush Nayak, Reecha Sahu, and Arunima Sur

Index *215*

Contributors

Latika Bhatia
Department of Microbiology and Bioinformatics, Atal Bihari Vajpayee University, Bilaspur, Chhattisgarh, India

Amandeep Brar
Department of Microbiology, Central University of Rajasthan, Ajmer, Rajasthan, India

Sushma Chauhan
Amity Institute of Biotechnology, Amity University Chhattisgarh, Raipur, Chhattisgarh, India

Ajay K. Dalai
Department of Chemical and Biological Engineering, University of Saskatchewan, Saskatoon, Saskatchewan, Canada

Janusz A. Kozinski
Faculty of Engineering, Lakehead University, Thunder Bay, Ontario, Canada

Tanushree Baldeo Madavi
Amity Institute of Biotechnology, Amity University Chhattisgarh, Raipur, Chhattisgarh, India

Mohit Mishra
Amity Institute of Biotechnology, Amity University Chhattisgarh, Raipur, Chhattisgarh, India

Vargobi Mukherjee
Amity Institute of Biotechnology, Amity University Chhattisgarh, Raipur, Chhattisgarh, India

Sonil Nanda
Department of Chemical and Biological Engineering, University of Saskatchewan, Saskatoon, Saskatchewan, Canada

Kush Nayak
Amity Institute of Biotechnology, Amity University Chhattisgarh, Raipur, Chhattisgarh, India

Sudheer D.V.N. Pamidimarri
Amity Institute of Biotechnology, Amity University Chhattisgarh, Raipur, Chhattisgarh, India

Nidhi Pareek
Department of Microbiology, Central University of Rajasthan, Ajmer, Rajasthan, India

Piyush Parkhey
Amity Institute of Biotechnology, Amity University Chhattisgarh, Raipur, Chhattisgarh, India

Chandramani Rai
Department of Chemical Engineering, Indian Institute of Technology Guwahati, Guwahati, Assam, India

Meenakshi Rajput
Department of Microbiology, Central University of Rajasthan, Ajmer, Rajasthan, India

Paresh H. Rana
Department of Chemical Engineering, Lalbhai Dalpatbhai College of Engineering, Ahmedabad, Gujarat, India

Reecha Sahu
Amity Institute of Biotechnology, Amity University Chhattisgarh, Raipur, Chhattisgarh, India

Prakash K. Sarangi
Directorate of Research, Central Agricultural University, Imphal, Manipur, India

Krushna Prasad Shadangi
Department of Chemical Engineering, Veer Surendra Sai University of Technology, Sambalpur, Odisha, India

Rakesh Kumar Soni
Department of Biotechnology, Shri Rawatpura Sarkar University, Raipur, Chhattisgarh, India

Arunima Sur
Amity Institute of Biotechnology, Amity University Chhattisgarh, Raipur, Chhattisgarh, India

Prabu Vairakannu
Department of Chemical Engineering, Indian Institute of Technology Guwahati, Guwahati, Assam, India

Balasubramanian Velramar
Amity Institute of Biotechnology, Amity University Chhattisgarh, Raipur, Chhattisgarh, India

Vivekanand Vivekanand
Centre for Energy and Environment, Malaviya National Institute of Technology, Jaipur, Rajasthan, India

Abbreviations

acetyl-CoA	acetyl coenzyme A
AEMs	anion exchange membranes
ATP	adenosine triphosphate
ATR	autothermal reforming
ATSR	autothermal steam reforming
BChl-a	bacterial chlorophyll-a
CE	coulombic efficiency
CEM	cationic exchange membrane
CFB	circulating fluidized bed
CLC	chemical looping combustion
CLG	chemical looping gasification
CLR	chemical looping reforming
CLSR	chemical looping steam reforming
CLWS	chemical looping water splitting
CMR	combined reforming of methane
CNTs-GF	CNT-modified graphite felt
COD	chemical oxygen demand
CP-AFM	conducting-probe atomic force microscopy
CPC	hierarchical porous carbon
DB	dry biomass
DBT	Department of Biotechnology
DFBR	dual fluidized bed reactor
DMR	dry reforming of methane
EAB	electrochemically active bacteria
ECR	electrochemical catalytic reforming
EET	extracellular electron transfer
ER	equivalence ratio
ETC	electron transport chain
FBR	fluidized bed reactor
Fd	ferredoxin
Fd_{red}	reduced ferredoxin
FETs	field-effect transistors
FHL	formate hydrogen lyase
HDPE	high-density polyethylene

HV	heating value
LDPE	low-density polyethylene
LHSV	liquid hourly space velocity
LHV	low-lower heating value
MDC	microbial dialysis cells
MEC	microbial electrolysis cell
MES	microbial electrochemical systems
MFC	microbial fuel cells
MSC	mesoporous silica carbon
MSW	municipal solid waste
NAD	nicotinamide adenine dinucleotide
NiFe LDH	nickel-iron layered double hydroxide
NiO	nickel oxide
NPs	nanoparticles
NREL	national renewable energy laboratory
OC	oxygen carrier
OEC	O_2 evolving complex
ORR	oxygen reduction reaction
OTM	oxygen transfer material
PC	phycocyanin
PCR	polymerase chain reaction
PE	phycoerythrin
PE	polyethylene
PEDOT	poly(3,4-ethylene dioxythiophene)
PEM	proton exchange membrane
PET	polyethylene terephthalate
PFL	pyruvate formate lyase
PFOR	pyruvate ferredoxin oxidoreductase
POX	partial oxidation
PP	polypropylene
PQ	plastoquinone
PS	polystyrene
PS-II	photosystem-II
PVC	polyurethane, polyvinyl chloride
RC-1	reaction center-1
RHA	rice husk ash
RHC	rice husk char
S/B	steam-to-biomass
S/C	stem-to-carbon

S/F	steam-to-fuel
SCWG	supercritical water gasification
SE	sorption enhanced
SE-CLSR	sorption enhanced chemical looping steam reforming
SESR	sorption enhanced steam reforming
SR	steam reforming
TG	thermogravimetric analysis
TG-FTIR	thermogravimetric-Fourier transform infrared spectroscopy
TMR	tri-reforming of methane
VFA	volatile fatty acids
WGS	water-gas shift

Preface

Renewable energy sources, such as solar, wind, tidal, geothermal, and biomass, are some promising alternatives to fossil fuels. Some of the examples of waste biomass resources are agricultural crop residues, forestry biomass, energy crops, invasive crops, cattle manure, sewage sludge, food waste, municipal solid waste, industrial effluents, and other organic residues. Different waste-to-energy technologies involving thermochemical, hydrothermal, physical/mechanical, biological, or integrated pathways can derive energy in the form of biofuels, bio-electricity, clean heat and power, and other bioenergy sources. Biofuels derived from waste biomass can be solid, liquid, or gaseous fuels. Biohydrogen is a promising gaseous biofuel with prospective applications in combined heat and power, fuel cells, or precursors for chemical production. Hydrogen can also be converted to liquid hydrocarbon fuels and value-added chemicals through catalytic thermochemical (e.g., Fischer-Tropsch synthesis) or biocatalytic biological (e.g., syngas fermentation) pathways. This book covers some recent advances in the production and utilization of biohydrogen. The book addresses the fundamentals and applied aspects of biohydrogen production, focusing on contemporary global research, emphasizing their technological, environmental, socio-economic, and techno-economic factors.

Chapter 1 by Rajput et al. emphasizes the basic principles, benefits, and challenges concerning both the biological and thermochemical routes for biohydrogen production. Chapter 2 by Rana deals with biomass conversion to hydrogen through gasification with a focus on the process parameters (e.g., biomass type, particle size and moisture content, reaction temperature, pressure, steam/biomass ratio, equivalence ratio, and catalyst-to-sorbent ratio), which considerably impact the syngas yield, composition, and fuel quality. Chapter 3 by Rai and Vairakannu comprehensively reviews the catalytic reforming technologies for hydrogen production concerning various feedstocks, conversion processes, hydrogen selectivity, and catalysis. Chapter 4 by Shadangi examines the prospects of co-conversion of plastic wastes and biomass into biohydrogen through co-gasification technology. The effect of process parameters (e.g., plastic-to-biomass ratio, steam-to-biomass ratio, equivalence ratio, temperature, gasifying agent, and catalysts) on syngas yields through co-gasification are vividly described in this chapter.

Chapter 5 by Sarangi et al. reviews fermentative hydrogen production technologies, primarily through photo-fermentation and dark fermentation. Chapter 6 by Chauhan et al. discusses the molecular mechanism of hydrogen production and enhances its yields by genetic and metabolic engineering of the native microbial host or installing the hydrogen machinery in non-hydrogen producing microbial host. Chapter 7 by Bhatia et al. reviews hydrogen production routes through microbial electrolysis, especially by microbial electrolysis cells and microbial fuel cells. Chapter 8 by Parkhey et al. discusses the electron transfer mechanism in electrochemically active bacteria and summarizes some notable studies conducted to identify novel electrode construction materials and assemblies for microbial electrochemical systems.

This book provides a comprehensive synopsis on the topics mentioned above for applications in several cross-disciplinary areas of biotechnology, fermentation technology, bioprocess engineering, catalysis, chemical engineering, and environmental sciences with a common agenda on biofuels and bioenergy.

The editors are grateful to all the authors for contributing their quality scholarly materials to develop this book. We express our sincere thanks to Ashish Kumar and Sandy Jones Sickels from Apple Academic Press, Inc. for their enthusiastic assistance while developing this book.

Dr. Sonil Nanda
Research Associate
Department of Chemical and Biological Engineering
University of Saskatchewan
Saskatoon, Saskatchewan, Canada
E-mail: sonil.nanda@outlook.com
ORCID: https://orcid.org/0000-0001-6047-0846

Dr. Prakash K. Sarangi
Scientist, Directorate of Research
Central Agricultural University
Imphal, Manipur, India
E-mail: sarangi77@yahoo.co.in
ORCID: https://orcid.org/0000-0003-2189-8828

CHAPTER 1

Recent Advancements in Biohydrogen Production: Thermochemical and Biological Conversion Routes

MEENAKSHI RAJPUT,[1] AMANDEEP BRAR,[1] VIVEKANAND VIVEKANAND,[2] and NIDHI PAREEK[1]

[1]*Department of Microbiology, Central University of Rajasthan, Ajmer, Rajasthan, India*
E-mail: nidhipareek@curaj.ac.in (Nidhi Pareek)

[2]*Center for Energy and Environment, Malaviya National Institute of Technology, Jaipur, Rajasthan, India*

ABSTRACT

To deal with the growing energy demand and climate change, carbon-neutral, renewable, and sustainable energy sources are highly essential. Biohydrogen has gained tremendous attention as a fuel due to its highest energy potential, and it holds great assurance as a sustainable fuel. Biohydrogen as a fuel does not generate any pollution and has the potential to reduce carbon emissions. Currently, biohydrogen is being produced from both biological and thermochemical conversion routes, and each route has its share of potential merits and demerits. This chapter emphasized the basic principles, benefits, and challenges concerning both the biological and thermochemical routes for biohydrogen production. Extensive measures should be taken to improve the yield of biohydrogen production in these systems and will further require large reactors to compensate for the low rate of biohydrogen production. Considerable research is the need of the hour for the improvement of biohydrogen production that can meet the requirement of the growing population.

1.1 INTRODUCTION

Biohydrogen holds great assurance as a dream fuel of the present and future generation as it holds many environmental and socio-economic benefits to its credit (Kotay and Das, 2008). The dependency on fossil fuels could be reduced via these alternate energy sources. Biohydrogen can reduce carbon emissions from different sectors. The current CO_2 levels have been exceeded by 412 parts per million by volume, and this increase has further elevated the greenhouse effect that has led to an increase in the global temperature (Chandrasekhar et al., 2015).

In the last decade, biohydrogen has gained tremendous interest from all the developed nations due to an increase in greenhouse gases, depleting fossil fuels, and the increasing cost of non-renewable sources. Furthermore, biohydrogen has the potential for the elimination of major problems that are being aroused by the use of fossil fuels. Moreover, 50 million tons of hydrogen are being traded annually throughout the world, having a 10% growth rate every year. It is well understood that the energy content of molecular hydrogen is 143 GJ/ton per unit weight, which is the highest among all the known gaseous fuels to date. It is almost 2.8-fold higher than the conventional hydrocarbon fuels (Boyles, 1984). Biohydrogen is the only known fuel that is carbon-free and on combustion oxidizes to water. Biohydrogen is being portrayed as the potential energy carrier of the future generation (Johnston et al., 2005).

At this moment, the production of hydrogen worldwide exceeds 1 billion m^3 per day and out of which the contribution of natural gas is 48%, the contribution of oil is 30% for hydrogen production, coal contributes 18% and the remaining 4% has been contributed by the electrolysis through the splitting of water. Biohydrogen can be used by direct combustion in an internal combustion engine as a fuel or a fuel cell (Kotay and Das, 2008). Fertilizers and petroleum industries are the largest user of hydrogen fuel with 50% and 37% share respectively. It is not only being used as an energy source but also as a raw material for the chemical production, production of devices for electronics, the hydrogenation of fats and oil in the food industry, in the processing of steel and refineries for desulfurization or re-formulation of gasoline. It is also used as a starting material for the production of alcohols, aldehydes, and ammonia.

With the emission-related standards getting strict day by day, the demand for clean fuel in the form of biohydrogen has increased drastically in the last ten years, and its biggest emerged user is the refineries. Hydrogen is being considered an affordable and domestic energy source that is safe and flexible for its use in all the economic sectors. In the coming decade, hydrogen has the capacity for becoming this world's clean energy source with electricity as the first carrier of energy and this will provide a strong foundation to the substantial energy system (Show et al., 2012). To date, most of the commercial production of hydrogen is through thermochemical processes such as reforming (Shanmugam et al., 2020). However, alternative clean processes for hydrogen production should be explored to reduce environmental impacts. The production of hydrogen from fuel cells and related technologies provides an essential link between renewable and sustainable energy sources (Levin et al., 2004).

1.2 AN OVERVIEW OF BIOHYDROGEN PRODUCTION METHODS

Biohydrogen has been considered as a burgeoning environment-friendly, a clean energy resource that yields water upon combustion. Comprehensively, it has been estimated that for the production of hydrogen, up to 96% of the production technologies rely on fossil fuels (Arregi et al., 2018). Thus, to satisfy the inordinately rising demand for energy while keeping in balance with fossil fuel consumption and CO_2 release reduction target is a major challenge. Thus, there is an urgent need to opt for a sustainable energy source such as biomass.

Biomass is recognized as the major renewable energy reserve globally, that can be used as a clean feedstock. Being a photosynthetic product, biomass is a plentiful and carbon-neutral renewable resource of energy that is capable of sustainable hydrogen production. Biomass receives significant attention as an energy source as besides its renewable and versatile nature it contains an appreciable amount of hydrogen, oxygen, and carbon. There are several types of biomass depending on their sources, including residual materials from agriculture and industries, municipal solid waste and sewage sludge (Figure 1.1) (Kaur et al., 2019). Consequently, for the production of biohydrogen, the biomass can be processed via the thermochemical route and biological route. The thermochemical route includes pyrolysis and gasification of biomass, whereas the biological route comprises microbes aided fermentation and photosynthesis (Figure 1.2). The following sections elucidate the production methods of biohydrogen from the biomass via the thermochemical route and biological route, respectively.

FIGURE 1.1 Types of biomass depending on their sources.

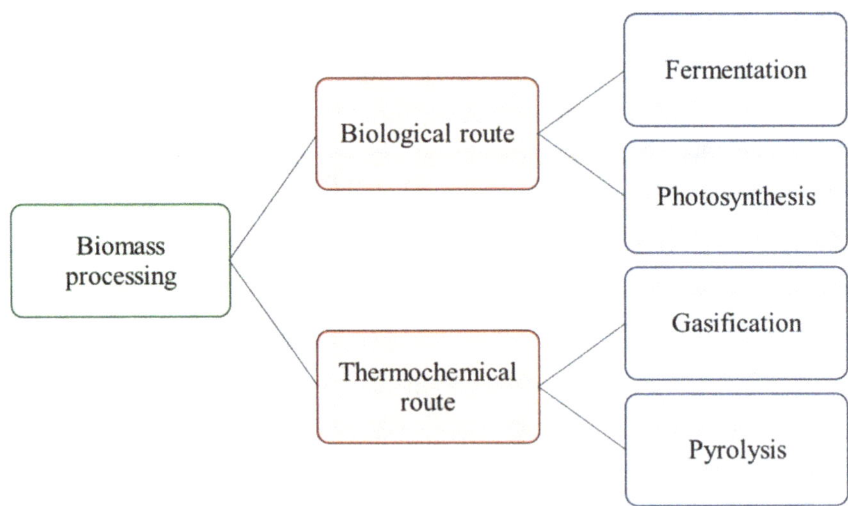

FIGURE 1.2 Biomass conversion methods to produce biohydrogen.

1.3 BIOHYDROGEN PRODUCTION VIA THERMOCHEMICAL CONVERSION ROUTE

As mentioned previously, hydrogen production can be achieved through the thermochemical processing of the biomass. The thermochemical conversion of biomass as feedstock yields bio-oil, water, tars, chars, and non-condensable gases (i.e., H_2, CH_4, CO_2, and CO). The formation of these reaction products is controlled by various parameters that involve heat source, the reactor used, temperature, heating rate and reaction environment (Kaur et al., 2019). The conversion of biomass as feedstock through the thermochemical route possesses several benefits, which include: (i) usage of whole biomass, (ii) less cost and energy input as pretreatment of biomass is not required, and (iii) short-term process. However, the major hurdle that comes in the commercialization of the process at an industrial scale is the yearlong constant availability of biomass concerning both quantity and quality (Bhaskar et al., 2013).

Several methods are employed for the thermochemical conversion of biomass comprising pyrolysis, gasification, combustion, and liquefaction. Pyrolysis and gasification are primarily used methods for the processing of biomass to generate hydrogen. Gasification is the process of partial oxidation that leads to the generation of a gaseous product, a concoction of H_2, CO_2, CO, and CH_4. The gasification of biomass is generally operated at 800–900°C which is followed by the reforming of the synthesis gas (H_2 + CO) (McKendry, 2002). At the same time, pyrolysis is known as the fundamental thermochemical process that decomposes the biomass to produce fuel gas along with solid char and bio-oil under oxygen-deprived conditions. Biomass conversion into hydrogen is aided largely by water-gas shift (WGS) reaction through gasification and pyrolysis. Furthermore, the purification of the hydrogen produced is done by the pressure swing adsorption process. The further sections rationalize the processing of biomass to generate biohydrogen via gasification and pyrolysis. The illustrative methods of hydrogen production from biomass are highlighted in Figure 1.3.

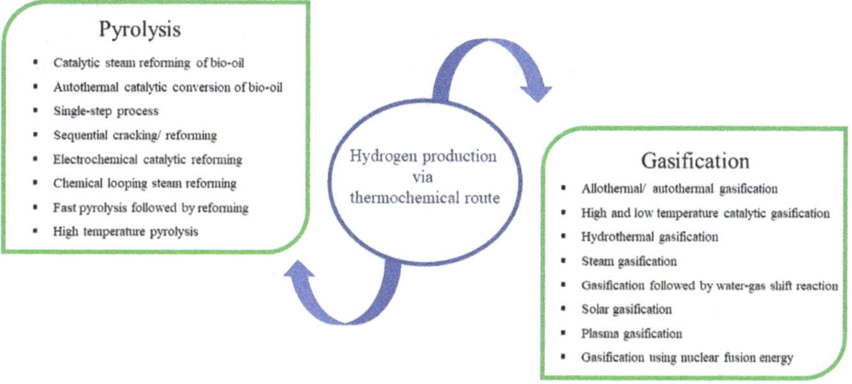

FIGURE 1.3 Different methods of hydrogen production via thermochemical routes.

1.3.1 *GASIFICATION*

Biomass gasification engages the thermo-chemical processing into the combustible gaseous product (a mixture of H_2, CO_2, CO, and CH_4) employing a gasifying agent (e.g., steam, air, and CO_2) at the relevant reaction conditions which involve high temperature (500–1400°C) and pressure (33 bar). Gasification is regarded as the most convenient and favorable process for the generation of energy through a partial or complete transformation of solid fuel sources into gases. Furthermore, the gasification technologies offer the opportunity for the sustainable conversion of biomass into highly desirable clean fuel gases or synthesis gas (syngas). The syngas is mainly composed of H_2 and CO (Demirbas, 2002; Maschio et al., 1994).

During biomass gasification, the smaller polymeric molecules are produced due to the breakdown of large polymers at the high-temperature range with the availability of an oxidizing agent. The smaller molecules further convert into permanent gases, ash, char, tar, and minor contaminants. Char and tar are generated via the incomplete decomposition of biomass. The hydrogen content of the reaction product varies from 15% to 20% when the air is used as a gasifying agent, which then rises to 30% to 40% in the case of the steam. The relative amount of CO_2, H_2, CO, CH_4, and H_2O produced in the gasification process relies on the process stoichiometry. For instance, when air is used as a gasifying agent, then N_2 contributes approximately half a fraction of the total gas produced. The air-to-fuel ratio varies between 0.2–0.35 during gasification; while in the case of steam, the ratio of steam-to-biomass is about 1.

The biomass gasification embodies a complex chemical process. It is the combination of both pyrolysis and gasification. The reaction begins with the pyrolysis step that brings about the thermal decomposition of biomass. This is an endothermic step yielding about 75–90% of volatile gaseous and liquid hydrocarbon, thus named as devolatilization step. The residual non-volatile substances having a high amount of carbon are termed as char. Subsequently, the conversion of char and volatile hydrocarbons leads to the production of syngas via gasification (Bridgwater and Evans, 1993). A couple of reactions involved in this step are as follows (Bridgwater and Evans, 1993; Bhaskar et al., 2013:

Exothermic Reactions:

Combustion: Char/biomass volatiles + $O_2 \rightarrow CO_2$ (1.1)

Partial Oxidation: Char/biomass volatiles + $O_2 \rightarrow CO$ (1.2)

Methanation: Char/biomass volatiles + $H_2 \rightarrow CH_4$ (1.3)

Water-Gas Shift Reaction: $CO + H_2O \rightarrow CO_2 + H_2$ (1.4)

CO Methanation Reaction: $CO + 3H_2 \rightarrow CH_2 + H_2O$ (1.5)

Endothermic Reactions:

Steam-Carbon Reaction:
Char/biomass volatiles + $H_2O \rightarrow CO + H_2$ (1.6)

Boudouard Reaction:
Char/biomass volatiles + $CO_2 \rightarrow 2CO$ (1.7)

Gasification is capable of attaining over 97% conversion of biomass into valuable syngas with no or greatly reduced production of unwholesome products such as oils, tars, char, or other undesirable wastes. This process also holds the potential to exploit the various types of biomass that differ in the degree of moisture content, extending from organic sludge with high moisture content to bone-dry wood. Unlike the conventional partial oxidation processes where even a little moisture content in the feedstocks leads to the alteration in energy efficiency, the gasification process employs more or less a fraction of this moisture content into the production of hydrogen and other fuel gases (Demirbas, 2009). Some other methods of biohydrogen production through biomass gasification are explained further.

1.3.2 BIOMASS GASIFICATION: ALLOTHERMAL AND AUTOTHERMAL

Gasification proceeds via an endothermic route, which implies that thermal energy (heat) is required to assist the process of gasification. Thus, based on the heat source, it could be either allothermal or autothermal. Allothermal gasification involves the transfer of heat necessary for the process from any external source such as band heaters, heating elements, and heat exchangers. For example, in coal gasification, the heat is provided to the coal from an external source, which means it is an allothermal gasification process. On the contrary, for autothermal gasification, the required heat for the process is generated inside the gasifier directly by partial oxidation. The valorization of oil palm agro-waste makes a perfect example of autothermal gasification as in this palm kernel shells are used for the heating purpose.

Allothermal and autothermal gasification are the two most essential technologies for the production of syngas. The syngas with Low-lower heating value (LHV) can be produced through autothermal gasification using air as the oxidizing agent, whereas medium-LHV syngas can be produced via allothermal gasification or steam/oxygen blown autothermal process. The autothermal and allothermal gasification of biomass is researched widely and a considerable amount of hydrogen production is also reported in various studies (Hu et al., 2016; Pio et al., 2017; Sun and Wu, 2019; Boujjat et al., 2020).

In allothermal gasifiers, the heat transfer from an external source can be carried out either by fitting heat exchangers inside the gasifiers or through circulated hot bed material for the efficient transfer of heat between the zone combustion and gasification. Two types of separate reactors are used for this process, which includes: (i) a gasifier where the conversion of biomass takes place into medium-heating value gas and residual char, and (ii) a combustor where the combustion of residual char occurs to produce heat for the gasification process. The circulating solid bed material (usually sand) present between the two reactors helps in heat transmission (Bhaskar et al., 2013). Rasmussen and Aryal (2020) reported a study on the syngas production through allothermal gasification of straw as feedstock and compared it with wood pellet. The straw has been gasified at 750, 800, 850, 900, and 950°C of furnace temperatures. They observed that at high allothermal temperatures significantly high production of CO, CH_4, and H_2 has been achieved, which was comparable to the wood pellet.

Autothermal reactors are generally inexpensive to construct, as the autothermal gasification process is capable of producing a sufficient amount of heat to withstand reaction conditions, which cuts the cost of mounting heat transfer area, heating elements and heat exchangers inside the gasifiers. However, sometimes working with these reactors get problematic because of the absence of temperature control through a cooling medium. Thus, the temperature inside the reactor can only be managed by adding a diluent or one of the components in excess amount. The most suitable autothermal reactor for biomass gasification is the Foster Wheeler-pressurized stem/ oxygen-blown circulating fluidized bed gasifier (Bhaskar et al., 2013).

1.3.3 BIOMASS GASIFICATION: HIGH AND LOW-TEMPERATURE CATALYTIC PROCESSES

Gasification of biomass in the existence of homogeneous and heterogeneous catalysts at a low range of temperatures between 350°C and 600°C, is called low-temperature catalytic biomass gasification. In contrast, when the biomass is gasified at high temperatures (500–700°C) with the aid of a catalyst, then this method is termed high-temperature catalytic biomass gasification (Kaur et al., 2019). At the high range of temperatures, the efficiency of gasification is enhanced due to the complete utilization of biomass. The factors influencing the performance of the process such as catalysts, the interaction between compounds and reaction mechanisms have been studied comprehensively in the literature.

Minowa et al. (1998) demonstrated the role of catalyst nickel and sodium carbonate while gasifying the cellulose at a low-temperature range (200–350°C). They concluded that sodium carbonate as a catalyst prevented the char generation from oil that eventually increased the total yield of oil whereas, nickel participated as a catalyst in the steam reforming reaction and methanation reaction. Supercritical water gasification of the organosolv-cellulose and lignin was performed using Ru/TiO_2 at 400°C (Osada et al., 2004). They further mentioned that a brown solid product is produced in the lack of a catalyst. Conversely, the existence of the catalyst (i.e., NaOH or Ni) led to the generation of a solid product with a comparatively greater yield of H_2. A high yield of CH_4 with the minimal solid product was obtained when ruthenium was employed as a catalyst. Consequently, formaldehyde, which acts as a model compound in biomass conversion for cross-linked agents, gets decomposed in the presence of a catalyst (Ru/TiO_2) and resulted in the generation of CH_4, CO_2, and H_2. However, in the absence of catalysts, it was decomposed into methanol and CO_2. Thus, indicating the crucial role played by a catalyst in the gasification of biomass at low temperature.

1.3.4 HYDROTHERMAL GASIFICATION

Hydrothermal gasification involves the utilization of sub and supercritical water during biomass gasification. The decomposition of biomass is accelerated due to the special thermophysical properties of subcritical and supercritical as green solvents (Reddy et al., 2014; Nanda et al., 2019a). Eventually, the degradation of large polymeric molecules of biomass gets fast, which further enhances the rate of the consecutive reactions that lead to the formation of gas at the low range of temperatures (Gong et al., 2017). In the hydrothermal gasification, water and other carbon-based (organic) solvents are used at the place of catalysts. The generation of tar and coke is immensely decreased in the hydrothermal gasification process as the reactive species developing while biomass decomposition is reduced by solvation in water that subsequently alleviates the polymerization rate (Kruse, 2009; Nanda et al., 2019b).

When the biomass is gasified in the water having its temperature between its boiling point to the critical temperature, i.e., 374°C, then the process is called subcritical hydrothermal gasification (Okolie et al., 2019). Many studies have been reported on the subcritical conversion of biomass employing different biomass as feedstocks. Several catalysts are also used in the process to enhance the production of syngas. On the other hand, supercritical water gasification involves water at its supercritical conditions (temperature > 374°C, pressure > 22.1 MPa) as a gasifying agent (Azargohar et al., 2019).

The function of catalyst (Ni) impregnated lignocellulosic biomass (wheat straw and pinewood) under sub- and supercritical conditions during hydrothermal gasification has been studied by Nanda et al. (2016b). They noticed that in comparison to the non-catalyst drenched biomass, Ni-impregnated biomass has resulted in significantly higher total gases and H_2 yields, i.e., 9.5–16.2 mmol/g and 2.8–5.8 mmol/g, respectively. The efficiency of carbon gasification was found to be 19.6–32.6%. Additionally, they reported that high hydrogen yields were achieved at a residence time of 45 min, 500°C and the biomass-to-water ratio of 1:10. Thus, it was concluded that nickel acted as a nano-catalyst by creating active sites in the lignocellulosic matrix for enhanced hydrogen production in supercritical water.

Comparative research of hydrothermal gasification under sub and supercritical conditions has been conducted between two different feedstocks, i.e., petroleum cake and asphaltene. Ni-impregnated activated carbon has been used as a catalyst to elevate the yields of syngas at premeditated optimal conditions (temperature 650°C, 15% concentration of feed and reaction time of 60 min). Particularly, high H_2 yield was obtained in supercritical hydrothermal gasification under optimal reaction conditions as 2.98 and 4.17 mmol/g for petroleum coke and asphaltene, respectively (Rana et al., 2019). In similar conditions, yields of CH_4 have also been calculated for petroleum coke and asphaltene as 1.07 and 2.54 mmol/g, respectively. The findings reveal that asphaltene serves as an outstanding feedstock to generate highly aromatic chars via hydrothermal gasification (Rana et al., 2019).

Salimi et al. (2016) have demonstrated the conversion efficiency of various agriculture residues including shells of almond and walnut, straw of barley, rice, wheat, and canola stalk to hydrogen through hydrothermal gasification using supercritical water as the solvent. Furthermore, they also studied the relation of lignin and cellulose present in biomass with the hydrogen yield. They disclosed that initial cellulose content in the feedstock is directly proportional to the total yield of gases obtained, while an inverse relation was observed between lignin content and total gas yields. The highest hydrogen yield (8.38 mmol/g) has been obtained from barley and canola stalk having the highest carbon/hydrogen ratio displayed 25.3 mmol/g of total gas yield. Thus, it can be inferred that a high yield of hydrogen can be obtained through subcritical and supercritical hydrothermal gasification.

1.3.5 STEAM ASSISTED BIOMASS GASIFICATION

The process of biomass gasification performed using steam as a gasifying agent for hydrogen production is referred to as steam gasification. The higher yield of hydrogen is obtained when steam is used at the place of air or CO_2 due to the decomposition of H_2O into additional H_2 molecules. Moreover, the higher heating value gas is obtained by using steam as compared to the partial oxidation using air because the dilution with N_2 is prevented. The following reaction takes place in this process (Kaur et al., 2019):

$$\text{Biomass} + \text{steam} \rightarrow H_2 + CH_4 + CO_2 + CO + \text{hydrocarbons} + \text{Char} + \text{tar} + H_2O \quad (1.8)$$

The process of steam gasification comprises of three stages that differ in reaction temperature (Pinto et al., 2003):

i. Devolatilization, maximum biomass (70–90%) gets transformed into volatile substances and char at relatively low-temperature (300–500°C);
ii. Cracking and reformation of volatile substances and char occur at a temperature greater than 600°C (Abatzoglou et al., 2000); and
iii. Char gasification, which takes place at the high range of temperatures greater than 800°C.

The mechanism of the reaction of this process involves a collection of complex competing reactions inclusive of gas-solid reactions occurring between the steam and biomass fuel particles and gas-gas reactions in which reactants are steam and the evolved gas species. The products obtained from these reactions include char, tar, and a product gas (H_2, CH_4, C_2H_4, CO, CO_2, and C_2H_6). The final composition of the product gas obtained during the biomass gasification is determined by the cracking and reforming reactions.

Xiao et al. (2017) reported a product gas with 41 vol% H_2 concentration, 14 g/Nm^3 tar content and 1 Nm^3/kg dry has yield produced by steam gasification of pine sawdust using an innovative decoupled dual loop gasification system. This system consisted of three reactors such as fuel reactor, reformer, and combustor that are used for pyrolysis and gasification, tar cracking and reforming as well as char combustion, respectively. Observations of this study indicate that a decoupled dual loop gasification system is an effective system to attain highly reduced tar formation.

1.3.6 BIOMASS GASIFICATION BY THE WATER-GAS SHIFT REACTION

The WGS reaction plays a vital role in the generation of hydrogen (H_2) while removing carbon monoxide (CO) in various energy-related chemical processes (Yao et al., 2017). The WGS reaction coupled with oxygen or air gasification is the most widely employed approach for biomass conversion into hydrogen. During the WGS reaction, CO_2 and H_2 are produced because of the reaction between H_2O and CO (1:1 molar ratio) on the active site of metal catalysts. The WGS reaction is shown as:

$$CO + H_2O \rightarrow H_2 + CO_2 \tag{1.9}$$

The WGS reaction is reversible, thus to shift the equilibrium towards the product side excess of steam is added. The WGS reaction is generally carried out at two different range of temperatures, which includes (Bartholomew and Farrauto 2006):

i. The high-temperature reactions proceed typically at 350–500°C in the presence of Fe and/or Cr-based catalyst; and
ii. The low-temperature reactions proceed typically at 200–250°C with Cu-Zn oxide as catalysts.

1.3.7 SOLAR GASIFICATION OF BIOMASS

Gasification employing solar energy as a heat source to generate hydrogen is termed solar gasification. Chen et al. (2010) reported a novel way of generating hydrogen from biomass by supercritical water gasification using solar energy (concentrated). The proof of concept for this novel system has been demonstrated by using glucose as model compounds and real biomass (i.e., wheat stalk and cornmeal) as feedstocks. The results showed high gasification efficiency with a 50% fraction of hydrogen. Steam reforming coupled with solar gasification has also proven to be an outstanding process for hydrogen production.

The integration of solar thermal energy (middle temperature) and methanol steam reformation has resulted in 2.6–2.9 mol as maximum hydrogen yield per mole of methanol. The experiment has been done at 150–300°C using a solar reactor of 5 kW under atmospheric pressure (Liu et al., 2009). The solar energy-driven gasification of biomass (wood of pine and spruce) was endeavored using combined drop tube or packed bed type solar reactor systems (1 kW). They also investigated the effect of various factors viz. temperature (1000–1400°C), biomass type and an oxidizing agent (CO_2 and H_2O) on the product gas. A linear increase has been observed in the hydrogen yield with temperature. Furthermore, the production of CH_4 showed an inverse relation with the secondary hydrocarbons (C_2H_2), while the yield of hydrogen was enhanced with the increase of secondary hydrocarbons (Bellouard et al., 2017).

1.3.8 PLASMA MEDIATED BIOMASS GASIFICATION

Gasification is performed at 1400°C using plasma, the process is referred to as plasma gasification or plasma arc decomposition. An electric arc present in the plasma torch helps in gas ionization and catalyzes the conversion of feedstock into syngas and slag. The plasma aids the process of biomass conversion into syngas by playing two roles, i.e., it decreases the production of tars and aerosols by acting as a purifying stage with simultaneous production of syngas rich in hydrogen (Bhaskar et al., 2013).

The significant heating of electrodes with subsequent cooling is the thermal plasma reforming process. When the hydrocarbons are not heated at high temperatures, a process called non-thermal plasma process. It is classified further as microwave plasma, corona effect, gliding arc discharge plasma and dielectric barrier discharge. Among all these processes, gliding arc discharge plasma is recognized as a commanding process as it reduced the development of tars and aerosols and produces valuable materials from slag (Kaur et al., 2019). The syngas produced in this process has a high concentration of both H_2 and CO while a low level of CO_2.

1.3.9 BIOMASS GASIFICATION USING NUCLEAR FUSION ENERGY

The process employs the steam produced from the nuclear power systems having high temperatures to achieve gasification. The process is used owing to its potentially high proficiency in hydrogen production than other processes. The hydrogen production from thermal decomposition of biomass (cellulose) with the aid of high-temperature steam released by nuclear fusion reactor has been demonstrated by Takeuchi et al. (2007). They concluded that this conceptual design for hydrogen production has the potential to produce more electricity and hydrogen as compared to the water or steam electrolysis.

1.3.10 PYROLYSIS

Pyrolysis is the thermal decomposition of biomass feedstock, which is operated at high temperature (375–525°C) and low pressure (0.1–0.5 MPa) in the absence of oxygen. In the combustion and gasification process, pyrolysis is the principal chemical reaction that also acts as a forerunner. The formation of liquid, gas or solid product is highly influenced by the reaction parameters such as residence time, rate of heating and temperature range (Bakhtyari et al., 2018). For instance, the generation of the gaseous product can be achieved through biomass pyrolysis where the temperature range is 1000–3000°C, the rate of heating is high and the residence time is less, whereas the formation of the liquid product is evident at the low-temperature range (Klass, 1998; Huber et al., 2006; Huber and Dumesic, 2006). The carbonaceous solid product (char) is generated when the residence time is comparatively high. However, the formation of a solid product or biochar during pyrolysis reduces the total yields of gaseous products and bio-oil (Nanda et al., 2016a). It is an endothermic reaction:

$$\text{Biomass} + \text{Energy} \rightarrow \text{Gas} + \text{Char} + \text{Bio-oil} \quad (1.10)$$

$$\text{Bio-oil} + H_2O \rightarrow H_2 + CO \quad (1.11)$$

$$CO + H_2O \rightarrow H_2 + CO_2 \quad (1.12)$$

Based on the reaction parameters (i.e., residence time, temperature, and rate of heating), pyrolysis is categorized in different categories as follows viz., slow, fast, and flash pyrolysis. The flash and fast pyrolysis mostly produce a liquid or gaseous products as they operate on less residence time and high temperature. Furthermore, the appropriate selection of biomass feedstock, reactor type, and heat transfer pattern is undeniably important to get the higher yields of desirable products. The operational reactors configurations that are vital for the pyrolysis include ablative, fluidized beds (circulating bubbling), vacuum, rotating cones and entrained flow. Additionally, the fluidized bed reactors and rotating cone reactors are highly suitable for pyrolysis as they are less expensive, easier for scaling, and possess high heating rates, which are essential for the process.

In the past few decades, the pyrolysis process has attracted much attention due to its potential of producing a greater amount of liquid fuel and cost-effectiveness. The pyrolysis oil obtained from hydrolysis of biomass owns a variety of applications like (i) direct utilization of fuel present in liquid form, (ii) serves as feed material in petrol-based industries and refineries, and (iii) improved to transport-level-fuel with some catalytic reactions (Kaur et al., 2019). The different methods of pyrolysis for biohydrogen production are displayed in Figure 1.3.

1.3.11 CATALYST-BASED STEAM TRANSFORMATION OF BIO-OIL

Hydrogen production by bio-oil could be best performed employing catalytic steam reformation using fluidized and fixed bed tubular reactors. A generalized way for Steam reforming of oxygenates is:

$$C_nH_mO_p + (n - p)H_2O \rightarrow nCO + (n - p + m/2)H_2 \tag{1.14}$$

The low temperature WGS reaction with the produced CO can recover more hydrogen.

$$nCO + nH_2O \rightarrow nCO_2 + nH_2 \tag{1.15}$$

The WGS reaction is an exothermic process that leads to the rise in temperature of the reactor, which further reduces the H_2 yield. The overall reaction could be shown as:

$$C_nH_mO_p + (2n - p)H_2O \rightarrow nCO_2 + (2n - p + m/2)H_2 \tag{1.16}$$

Steam reforming undergoes endothermic reactions that increase the temperature of the reactor that in turn results in the decomposition reactions, which further lead to the generation of carbonaceous coke that blocks the reactor and the activity of the catalyst. This is the major drawback of this approach because the deposition of coke over catalyst deactivates it, which mandates the need for frequent regeneration of the catalyst.

A comparative study for hydrogen production was done in between fixed- and fluidized bed reactors employing Ni/MgO-La_2O_3-Al_2O_3 as a catalyst. The catalyst was formed using nickel as an active agent, MgO, and La_2O_3 as promoters and Al_2O_3 as support. The reaction parameters including temperature, steam-to-carbon (S/C) ratio and liquid hourly space velocity (LHSV) were examined in terms of hydrogen production for both the reactors. The results disclosed that the fluidized bed reactor obtained 7% more hydrogen yield as compared to the fixed bed reactor. In the fluidized bed reactor, the highest hydrogen yield obtained was about 75.9% at temperature 700–800°C, S/C ratio 15–20 and LHSV 0.5–1.0 h^{-1}. Furthermore, the deposition of carbon was less on a fluidized bed reactor that rendered hydrogen generation in a fixed bed reactor (Lan et al., 2010).

Wang et al. (2014) prepared a nickel/ash catalyst by the process of wet impregnation using solid waste coal ash as support. The researchers reported and 83.5% and 98.4% conversion of phenol and acetic acid, respectively at temperature 700°C and S/C ratio 9.2. The total hydrogen yield of 79.1% and 85.6% were observed for phenol and acetic acid, respectively. Furthermore, the Ni/ash catalyst was stable for more than 10 h. The findings signify the comparable performance of the Ni/ash catalyst with the Ni-based commercial catalyst.

1.3.12 AUTOTHERMAL CATALYTIC TRANSFORMATION OF BIO-OIL

The process has been considered as an alternate for the generation of hydrogen via bio-oil steam reforming. The National Renewable Energy Laboratory (NREL) has set up a new method for hydrogen production from stabilized pyrolysis oil (stabilized with 10% methanol). The process proceeds through three steps:

i. Volatilization at temperature 400°C

ii. Oxidative cracking (650°C)

iii. Conversion through autothermal catalytic mode to yield hydrogen with less production of secondary and tertiary tar.

The separation of hydrogen from the mixture of the product obtained (i.e., H_2O, CO, CH_4, and CO_2) was achieved using a supported membrane, while the combustion of residual gas provided heat for the process. Autothermal steam reforming (ATSR) is used over conventional steam reforming because the thermal energy generated by the exothermic process of partial oxidation or complete oxidation of bio-oil or oxygenates reimburse the requirement of heat in endothermic reforming reactions (Kaur et al., 2019). The ATSR process is capable of minimizing the heat load demand in the reformer. Czernik and French (2014) illustrated the autothermal reforming of fast pyrolysis liquids obtained from three different biomass feedstock (i.e., oak, poplar, and pine) using a noble metal catalyst (0.5% Pt/Al_2O_3). The reformer temperature was set between 800–850°C. The results obtained showed a total of 9–11 g of H_2/100 g of bio-oil, which resembles the stoichiometric potential of 70–83%. Furthermore, the authors reported that the bio-oil composition and carbon-to-gas ratio influences the yield of H_2.

1.3.13 MONO-STEP PRODUCTION OF HYDROGEN-RICH GAS

The process involves the direct hydrogen generation from biomass pyrolysis impregnated with certain metal salts. The biomass lignin, cellulose, and hemicellulose (i.e., beechwood xylan) impregnated with nickel and iron solutions were dried and pyrolyzed (600°C). The beechwood samples were also impregnated with different concertation of metal content to evaluate the role of Ni and Fe in the pyrolysis of biomass. The results obtained signified that due to the metal impregnation, considerable changes have been observed in the tar concentration (gas chromatography-mass spectrometry analysis), product yield and composition of the gaseous fraction. Furthermore, Ni has proven to be more efficient than Fe owing to its ability to rearrange the aromatic rings present in the matrix, which consequently increased the production of H_2 and decreased the aromatic tar production (Collard et al., 2012).

Collard et al. (2015) investigated the influence of Ni and Fe impregnated cellulose on fast pyrolysis using coupled thermogravimetric analyzer and micro-gas chromatography techniques. Findings indicate that a few rearrangement reactions have been catalyzed by impregnated metal when the catalyst was loaded in fewer amounts. Furthermore, at 210°C Ni-impregnated cellulose showed a high conversion rate. Thus, it is evident that the impregnation of the metal catalyst with biomass has significantly increased the efficiency of conversion and hydrogen yield at low temperatures.

1.3.14 SEQUENTIAL CRACKING AND BIO-OIL REFORMING

The sequential cracking/reforming process involves two alternate steps of the cracking reaction step and the regeneration step. The cracking reaction involves the conversion of bio-oil into a gaseous mixture comprising of H_2, CH_4, CO, and CO_2 while the regeneration step leads to reactivation of catalyst via combustion of the deposited carbon on active sites of the catalyst. The process is desirable for generating hydrogen as the formation of H_2 and CO_2 takes place at alternate steps that further reduces the energy required for the purification of hydrogen produced.

A study was conducted on the conversion of crude bio-oil into hydrogen-rich gas using the sequential catalytic process. The process involves the alternate steps of cracking/reforming during which a stream rich in H_2 and CO was produced and the carbon deposited on the catalyst was combusted in the air during the regeneration step. They employed two Ni-based (i.e., Ni/Al_2O_3 and $Ni\text{-}K/La_2O_3\text{-}Al_2O_3$) catalysts in the process. The nickel-based catalyst efficiently catalyzed the hydrogen production using bio-oil yielding a gaseous stream comprising 40–50% of H_2 (Davidian et al., 2007).

1.3.15 ELECTRO-CHEMICAL CATALYTIC TRANSFORMATION OF BIO-OIL

Electrochemical catalytic reforming (ECR) is an excellent and potent approach for reforming bio-oil at low temperatures to yield more hydrogen. A study has been carried out to determine the efficiency of a novel ECR approach over a conventional nickel-based catalyst (i.e., NiO/Al_2O_3) approaches. Hydrogen yield and carbon conversion efficiency obtained through ECR was above 90% at even low temperatures. Thus, the study concludes that in the ECR process, thermal electrons aids the disintegration and reforming reaction of organic oxygenates present in bio-oil (Yuan et al., 2008). Ye et al. (2009) studied the hydrogen production employing a low-temperature ECR approach over nickel-based catalyst $NiCuZn/Al_2O_3$. They observed nearly complete conversion and about 93.5% of H_2 yield at low reforming temperature (400°C).

1.3.16 BIO-OIL REFORMING: CHEMICAL LOOPING

Chemical looping combustion (CLC) uses metal as an oxygen carrier, which is oxidized by air in a reactor named as the oxidizer and further in the second reactor get reduced in contact with fuel. The process of CLC in which steam plays the role of oxidant is referred to as chemical looping steam reforming (CLSR) (Solunke and Veser, 2010). The advantages of using the CLSR process include that it can be a steam reforming reaction that can be carried out at a low temperature, and problems of catalyst coking are also resolved.

Udomchoke et al. (2016) attempted to improvise the sorption enhanced (SE)-chemical looping reforming process (CLR) required to produce hydrogen after converting bio-oil derived from pyrolysis of corn stover. For this, they focused on the minimal energy usage via the thermal isolation process of the reformer, air reactor and calcination reactor used. They also modified the process of SE-CLR by incorporation a regeneration step for catalyst and sorbent. NiO and CaO-CO_2 were employed as the catalyst/oxidizing agent and sorbent, respectively. It was monitored that the total amount of bio-oil converted was 92% as well as the total amount of hydrogen yielded was 153.4 g in 1 kg corn stover. Furthermore, the bio-oil conversion efficiency was enhanced as more catalyst and sorbent (solids) were added and were fed to the reformer. Thus, the modification step of the recirculation of solids from the air reactor to the reformer directly has significantly enhanced the production of hydrogen along with the bio-oil conversion efficiency.

Dou et al. (2018) illustrated that the continuous highly purified oxygen was produced through the integration of oxidation, steam reformation, WGS reaction and CO_2 capture (*in situ*) in a fixed bed reactor. NiO/Al_2O_3 was used as an oxygen carrier (OC) and CaO as a sorbent to evaluate the influence of adding sorbent for CO_2 removals on hydrogen production from ethanol through the SE-CLSR process. Results obtained indicated that the reduction of OC by ethanol has played an imperative role in catalyst aided steam reforming and WGS for hydrogen production. On the other hand, CaO as sorbent removed the CO_2 proficiently which led to the intensification of the process.

1.3.17 FAST PYROLYSIS PRECEDING STEAM REFORMING

The organic materials are heated intensively without air for the production of organic vapors, pyrolysis gases and charcoal during fast pyrolysis. The condensation of the vapors leads to the production of bio-oil. Furthermore, the bio-oil undergoing steam reforming results in hydrogen generation. Bio-oil transportation is easy as well as economical as compared with organic material or hydrogen. Thus, biomass conversion into bio-oil and production of hydrogen is possible upon different locations that are fully augmented with the feedstock supply and hydrogen distribution infrastructure. Two bio-oil samples (C1 and F1) derived from the fast pyrolysis of rice husk were subjected to steam reforming at three different temperatures (650, 750, and 850°C) in a fixed bed reactor using a commercial catalyst. Gas chromatography and mass spectrometry techniques were used to determine the gas composition and organic composition, respectively in the liquid condensate. The highest hydrogen efficiency of 45.3% was obtained at 850°C when more water was added for F1 bio-oil. Furthermore, 55.8–61.3% was the estimated range of hydrogen mole fraction at all the temperatures. However, the carbon deposition hinders the conversion of bio-oil and it formed easily at 750°C (Chen et al., 2011).

1.3.18 BIOMASS PYROLYSIS AT HIGH TEMPERATURES

A process of rapid as well as transitional pyrolysis undergoing at temperatures ≥700°C results in the production of gas without the tar, which is referred to as high-temperature pyrolysis. This process can produce significantly high-quality product gas. The utilization of catalysts, *viz.* Ni and dolomites at the higher range of temperature along with the oxygen and steam could be useful in getting more gaseous products (Kaur et al., 2019). In a study, the sewage sludge was pyrolyzed using an electronic oven and microwave to produce a higher amount of gases, to evaluate the fineness to be used as fuel and to check the presence of hydrogen or syngas. The graphite and char were used as a microwave absorber. The Fourier-transform infrared (FTIR) analysis of pyrolysis oil showed that they are composed of hydrogen along with ester, carboxylic acid and amide groups. Moreover, the total fraction of hydrogen and syngas obtained was 98% and 66%, respectively. It can be concluded that high-temperature microwave pyrolysis is more efficient than conventional pyrolysis for the production of syngas and hydrogen (Domínguez et al., 2006).

1.4 BIOHYDROGEN PRODUCTION VIA BIOCHEMICAL CONVERSION RATE

The production of biohydrogen by biological means is one of the emerging technologies with sustainability and eco-friendliness. Biohydrogen production is less energy-intensive with biological processes as they occur at ambient temperatures and pressures when compared to hydrogen production from thermochemical processes (Nanda et al., 2017; Sarangi and Nanda, 2020).

Different taxonomic and physiological varieties of microorganisms are efficient in the production of biohydrogen. These microorganisms use the enzymes *viz.* hydrogenase and nitrogenase as hydrogen-producing proteins. These enzymes play a significant role in regulating the hydrogen metabolism in a range of prokaryotic and many eukaryotic microorganisms including green algae. These microorganisms involve the process of biophotolysis, which happens because of the light effect leading to the splitting of water into molecular oxygen and hydrogen (Azwar et al., 2014). The biological pathways that are involved in the synthesis of biohydrogen are either light-dependent or light-independent pathways.

1.4.1 LIGHT-DEPENDENT BIOSYNTHETIC PATHWAY OF HYDROGEN SYNTHESIS

This pathway consists of photolysis, both direct and indirect and photo-fermentation. In direct photolysis, the effect of light on the biological system leads to the dissociation of the substrate into water that further results in the production of hydrogen. For biohydrogen production, direct photolysis uses solar energy and the photosynthetic system of microorganisms (especially algae) for the conversion of water into chemical energy.

To study in this area has helped to decode the processes involved, such as molecular events that are associated with photosynthesis and the subsequent reactions that occur in the metabolism of microorganisms. The photosynthetic system consists of two photosystems that operates in the well-coordinated series and capture two quanta of light and cause splitting of the electron from the water molecule. For the complete dissociation of the oxygen molecule, eight quanta of light are required. To use the photosynthetic protons for the generation of hydrogen, the bacterial hydrogenase enzyme plays an important role. For this process, an electron acceptor molecule must have an oxidation-reduction potential equivalent to that of hydrogen. When the electron acceptor is in a reduced state, it acts as a substrate for the hydrogenase enzyme (Pandu and Joseph, 2012).

Oxygen synthesized during photosynthesis usually inhibits the enzyme hydrogenase and to overcome this, the photosynthetic process requires the production of oxygen absorbers in the form of glucose-oxidase. In atmospheric conditions, biohydrogen production occurs in two phases. In the first phase, cell growth takes place, and in the second phase, the evolution of hydrogen takes place. Direct photolysis depicts the working model of hydrogen production by using sunlight and water.

Indirect-photolysis, the major problem of the susceptibility of hydrogen production is prevented by the separation of oxygen and hydrogen generation. Therefore, this process involves the complete separation of evolution of hydrogen and oxygen at different stages that are being coupled by the CO_2 fixation (Manish and Banerjee, 2008). The classical example of this process is microalgal and cyanobacterial species. These species in the presence of sunlight, use six molecules each of water and CO_2 to form a six-carbon carbohydrate along with six molecules of oxygen. Then the six-carbon carbohydrate reacts with six molecules of water to yield 12 molecules of hydrogen and six molecules of CO_2.

In photo-fermentation, energy is being derived from the sunlight and the green algae release the electrons by photosynthesis for use in the endogenous catabolism of a substrate. In photo-fermentation, photosynthetic microorganisms convert the solar energy into biohydrogen by the use of organic molecules. This process produces a comparatively lower yield of biohydrogen in comparison to thermochemical routes, but this process has an advantage due to its enhanced efficiency of biohydrogen without the need for light and variety of feed (Show et al., 2012). Many photosynthetic bacteria have been explored for this process because they lack Photosystem II that eliminated the problem associated with the inhibition of hydrogenase enzyme by oxygen.

1.4.2 LIGHT-INDEPENDENT BIOSYNTHETIC PATHWAY OF HYDROGEN SYNTHESIS

This pathway is mainly represented by dark-fermentation. In this process, the anaerobic microorganisms yield hydrogen by the fermentation of organic substrates. Dark fermentation is a very complex process that involves a series of reactions enabled by a diverse group of microorganisms. In this process, the complex molecules are degraded into their subsequent monomeric units by the anaerobic microorganisms. The hydrogen-producing microorganisms (facultative or anaerobic) further degrade these monomers to low molecular weight organic acids and alcohols. These reactions occur at a wide range of temperatures from being mesophilic to hyperthermophilic.

The wastewater is being considered for the dark fermentation processes for biohydrogen production and the hydrogen generated is comparatively higher than the other methods. The wastewater is being used from different sources that are rich in organic substrates. The utilization of wastewater as a substrate diminishes the additional cost for biohydrogen production. In the dark fermentation, an anaerobic reaction forms hydrogen and methane in two steps. Hydrogen is formed in the first reaction as an intermediate product, and this is further used in the subsequent reaction of methanogenesis. Moreover, it results in the biodegradation of the wastewater. Hydrogen is generated by harvesting it in the first step and by inhibiting the second step where microbes consume the hydrogen for methanogenesis. This is being achieved by lowering the pH of the bioreactor or by inhibiting the hydrogen-producing microbes by adding various chemicals (Kim et al., 2005; Hallenbeck and Ghosh, 2009).

When hydrogen is used as fuel in engines, various problems such as its purification from a mixture of gases, its storage and transportation arise. The purification of hydrogen is the critical step for the employment of hydrogen as a fuel. The advantages of dark fermentation include comparatively the rate of hydrogen production is high, this process has low-energy requirements, its simple operation and the process is sustainable due to its efficiency to use wastewater as a substrate. Other substrates that can be used in dark-fermentation include lignocellulosic biomass, food waste, municipal solid waste, glucose, glycerol, etc. (Okolie et al., 2020).

1.4.3 FERMENTATION TECHNOLOGY

The hybrid fermentation technology based on both the dark and photo-fermentation might offer a beneficial route for biohydrogen yield. This relies on the complete utilization of substrate for the biohydrogen production, which is otherwise not possible due to thermochemical limitations. Therefore, in this hybrid technology photosynthetic and non-photosynthetic bacteria provides an integrated system for hydrogen production. Through anaerobic digestion, the complex substrate is degraded into simpler compounds that are being used by the photosynthetic bacteria for hydrogen production (Dasgupta et al., 2010). A few examples have been cited in Table 1.1.

TABLE 1.1 Biomass Used for Biohydrogen Production

Experimental Conditions	Findings	References
• *Chlorella* sp. was used to study energy recovery through a two-step fermentation process. • The fermentative bacteria were collected from heat-treated sludge.	• 8.29 ± 0.33 mL of H_2 per g of volatile solids. • Energy recovery increased by 22–146% in a two-stage reaction.	Wu et al. (2020)
• Swine manure as a substrate was fermented for biohydrogen production. • *Clostridium* was used in a fed-batch reactor.	• pH and hydraulic retention time have a direct effect on fermentative biohydrogen production. • 18.7 g of H_2 per g of total volatile solids.	Zhu et al. (2009)
• Biohydrogen was produced from the fermentation of biomass of *Spirogyra* sp. by the *Clostridium butyricum* DSM 10702. • The biomass subjected to nitrogen stress for carbohydrate induction.	• *Spirogyra* hydrolysate has shown hydrogen production of 3.9 L/L of H_2, which is equivalent to 146.3 mL H_2 per g dry weight of microalga.	Pinto et al. (2018)
• Enhancement in the biohydrogen production employing *Chlorella* sp. with thermal treatment through fermentation.	• 1079 mL of H_2/L with a yield of 54 mL H_2 per g of volatile solids • The treatment solubilizes the algal biomass	Giang et al. (2019)
• Hydrogen generation prospective of *Humulus scandens* through dark fermentation by *Enterobacter aerogenes*.	• H_2 production potential was 65.1 mL per g of total solids • Optimized cellulase amount, inoculation, and initial pH	Zhang et al. (2020)

1.5 OPPORTUNITIES AND CHALLENGES IN THE PRODUCTION OF BIOHYDROGEN

With the depleting resources of non-renewable energy sources, the focus has been shifted to hydrogen, as it is one of the most promising, eco-friendly, and sustainable energy sources. Despite having many advantages, the largescale production of biohydrogen is restricted due to many constraints, thereby affecting its commercial production (Shanmugam et al., 2020). There are many technical challenges in the large-scale production of biohydrogen, such as decreasing the production cost, reducing the supplying expenses along with its storage cost, conversion, and applications (Kotay and Das, 2008).

The production of biohydrogen from different sources requires abundant energy input in the form of thermal energy, electrolytic (electricity) or photolytic (light) energy. The major disadvantage that is associated with the biological system is the relatively low hydrogen yield and its production rate. Methods such as improvement in the microbial strains, troubleshooting of process modification, development of an efficient reactor design along with the application of genomics and proteomics can pave the way for the successful development of biohydrogen production systems (Guwy et al., 2011).

The hydrogen production through the thermochemical conversion of biomass has been recognized as one of the most propitious approaches for the utilization of biomass. At present, biomass is considered the major energy reserves for the world. Biomass has been regarded as the main source of concentrated carbon available for sustainable energy production. The primitive biomass conversion process, i.e., combustion was in practice for centuries as people used to burn the biomass in the open air to generate heat. However, the combustion is restricted now as it is not a favorable approach from the environmental point of view. Besides, thermochemical processes also hold a long scientific history as they came into practice to overcome the limitations of combustion.

The thermochemical methods such as pyrolysis and gasification followed by steam reforming have enormous potential to be executed at the industrial level as compared to other conventional thermochemical processes. Gasification of biomass offers the possibility to generate highly efficient clean energy at a large scale by exploiting biomass sources despite being relying on insecure fossil fuels. As compared with other thermochemical processes, the supercritical water gasification process is capable of directly using wet biomass even at a low-temperature range. However, the large heat input and high capital cost due to high operating pressure impart a negative impact on the economic performance of the supercritical water gasification process.

Another major issue that makes the process inefficient is the formation of a large quantity of carbonaceous compounds, i.e., char, and tar. However, for industrial implementation, these processes are still in their primitive stage, but it is expected that these thermochemical processes could become the most economical options for the supply of sustainable hydrogen in the coming years.

1.6 CONCLUSIONS

Biohydrogen is being termed as the energy source of the future generation that meets the requirements of a sustainable environment. Like fossil fuels and natural gas, it is not readily available in nature. Therefore, there is an urge to develop processes that are efficient and environment-friendly for sufficient production. The thermochemical methods include steam reforming, pyrolysis, and gasification, which might require high temperatures for operation. This requirement makes these processes energy-intensive and expensive.

The biological production of biohydrogen is a novel and promising approach that produces clean energy sources by using waste materials for meeting the energy needs of the growing generation. This production method requires mild operational conditions, thereby consuming less energy, but it requires raw materials that further enhance the cost of the process. By using solid waste that is rich in carbohydrates, starch or cellulose can reduce the overall cost of the process, thereby making the biological production of biohydrogen more attractive. Extensive measures should be taken to improve the yield of biohydrogen production in this system. This will further require large reactors to compensate for the low rate of biohydrogen production. Therefore, considerable research is the need of the hour for the improvement of biohydrogen production that can meet the need of the ever-growing population.

KEYWORDS

- **biofuels**
- **biohydrogen**
- **carbon monoxide**
- **hydrocarbons**
- **liquid hourly space velocity**
- **steam-to-carbon ratio**
- **thermochemical conversion**

REFERENCES

Abatzoglou, N., Barker, N., Hasler, P., & Knoef, H., (2000). The development of a draft protocol for the sampling and analysis of particulate and organic contaminants in the gas from small biomass gasifiers. *Biomass Bioenergy, 18*, 5–17.

Arregi, A., Amutio, M., Lopez, G., Bilbao, J., & Olazar, M., (2018). Evaluation of thermochemical routes for hydrogen production from biomass: A review. *Energy Convers Manag., 165*, 696–719.

Azargohar, R., Nanda, S., Dalai, A. K., & Kozinski, J. A., (2019). Physico-chemistry of biochars produced through steam gasification and hydrothermal gasification of canola hull and canola meal pellets. *Biomass Bioenergy, 120*, 458–470.

Azwar, M. Y., Hussain, M. A., & Abdul-Wahab, A. K., (2014). Development of biohydrogen production by photobiological, fermentation and electrochemical processes: A review. *Renew Sust. Energ. Rev., 31*, 158–173.

Bakhtyari, A., Makarem, M. A., & Rahimpour, M. R., (2018). Hydrogen production through pyrolysis. In: Encyclopedia of sustainability science and technology. In: Meyers, R., (ed.), *Fuel Cells and Hydrogen Production* (Vol. 2, p. 947). Springer New York.

Bartholomew, C. H., & Farrauto, R. J., (2006). *Fundamentals of Industrial Catalytic Processes: Second Edition* (2nd edn.). Wiley Online Library.

Bellouard, Q., Abanades, S., Rodat, S., & Dupassieux, N., (2017). Solar thermochemical gasification of wood biomass for syngas production in a high-temperature continuously fed tubular reactor. *Int. J. Hydrogen Energy, 42*, 13486–13497.

Bhaskar, T., Balagurumurthy, B., Singh, R., & Poddar, M. K., (2013). Thermochemical route for biohydrogen production. In: Pandey, A., Chang, J. S., Hallenbeck, P. C., & Larroche, C., (eds.), *Biohydrogen* (pp. 285–316). Elsevier, USA.

Boujjat, H., Yuki -Junior, G. M., Rodat, S., & Abanades, S., (2020). Dynamic simulation and control of solar biomass gasification for hydrogen-rich syngas production during allothermal and hybrid solar/autothermal operation. *Int. J. Hydrogen Energy., 1*–11.

Boyles, D. T., (1984). *Bio-energy: Technology, Thermodynamics, and Costs.* Chichester, West Sussex, Ellis Horwood.

Bridgwater, A. V., & Evans, G. D., (1993). *An Assessment of Thermochemical Conversion Systems for Processing Biomass and Refuse*. United Kingdom.

Chandrasekhar, K., Lee, Y. J., & Lee, D. W., (2015). Biohydrogen production: Strategies to improve process efficiency through microbial routes. *Int. J. Mol. Sci., 16*, 8266–8293.

Chen, J., Lu, Y., Guo, L., Zhang, X., & Xiao, P., (2010). Hydrogen production by biomass gasification in supercritical water using concentrated solar energy: System development and proof of concept. *Int. J. Hydrogen Energy, 35*, 7134–7141.

Chen, T., Wu, C., & Liu, R., (2011). Steam reforming of bio-oil from rice husks fast pyrolysis for hydrogen production. *Bioresour. Technol., 102*, 9236–9240.

Chen, X., Chen, Y., Yang, H., Chen, W., Wang, X. H., & Chen, H., (2017). Fast pyrolysis of cotton stalk biomass using calcium oxide. *Bioresour. Technol., 233*, 15–20.

Collard, F. X., Bensakhria, A., Drobek, M., Volle, G., & Blin, J., (2015). Influence of impregnated iron and nickel on the pyrolysis of cellulose. *Biomass Bioenergy., 80*, 52–62.

Collard, F. X., Blin, J., Bensakhria, A., & Valette, J., (2012). Influence of impregnated metal on the pyrolysis conversion of biomass constituents. *J. Anal. Appl. Pyrolysis, 95*, 213–226.

Czernik, S., & French, R., (2014). Distributed production of hydrogen by auto-thermal reforming of fast pyrolysis bio-oil. *Int. J. Hydrogen Energy, 39*, 744–750.

Dasgupta, C. N., Gilbert, J. J., Lindblad, P., Heidorn, T., Borgvang, S. A., & Das, S. K., (2010). Recent trends on the development of photobiological processes and photobioreactors for the improvement of hydrogen production. *Int. J. Hydrogen Energy, 35*, 10218–10238.

Davidian, T., Guilhaume, N., Iojoiu, E., Provendier, H., & Mirodatos, C., (2007). Hydrogen production from crude pyrolysis oil by a sequential catalytic process. *Appl. Catal. B: Environ., 73*, 116–127.

Demirbaş, A., (2002). Hydrogen production from biomass by the gasification process. *Energy Sources, 24*, 59–68.

Demirbas, A., (2009). Biohydrogen. In: *Green Energy and Technology*. Springer London.

Domínguez, A., Mene, J. A., Inguanzo, M., & Pıs, J. J., (2006). Production of bio-fuels by high temperature pyrolysis of sewage sludge using conventional and microwave heating. *Bioresour. Technol., 97*, 1185–1193.

Dou, B., Zhang, H., Cui, G., Cui, G., Wang, Z., Jiang, B., Wang, K., Chen, H., & Xu, Y., (2018). Hydrogen production by sorption-enhanced chemical looping steam reforming of ethanol in an alternating fixed-bed reactor: Sorbent to catalyst ratio dependencies. *Energy Convers. Manag., 155*, 243–252.

Giang, T. T., Lunprom, S., Liao, Q., Reungsang, A., & Salakkam, A., (2019). Improvement of hydrogen production from *Chlorella* sp. biomass by acid-thermal pre-treatment. *Peer. J., 21*, 6637.

Gong, M., Nanda, S., Romero, M. J., Zhu, W., & Kozinski, J. A., (2017). Subcritical and supercritical water gasification of humic acid as a model compound of humic substances in sewage sludge. *J. Supercrit. Fluids, 119*, 130–138.

Guwy, A. J., Dinsdale, R. M., Kim, J. R., Massanet-Nicolau, J., & Premier, G., (2011). Fermentative biohydrogen production systems integration. *Bioresour. Technol., 102*, 8534–8542.

Hallenbeck, P. C., & Ghosh, D., (2009). Advances in fermentative biohydrogen production: The way forward?. *Trends Biotechnol., 2*, 287–297.

Huber, G. W., & Dumesic, J. A., (2006). An overview of aqueous-phase catalytic processes for production of hydrogen and alkanes in a biorefinery. *Catal. Today, 111*, 119–132.

Huber, G. W., Iborra, S., & Corma, A., (2006). Synthesis of transportation fuels from biomass: Chemistry, catalysts, and engineering. *Chem. Rev., 106*, 4044–4098.

Johnston, M. D., Lawson, S., & Otter, J. A., (2005). Evaluation of hydrogen peroxide vapour as a method for the decontamination of surfaces contaminated with *Clostridium botulinum* spores. *J. Microbiol. Methods, 60*, 403–411.

Kaur, R., Gera, P., Jha, M. K., & Bhaskar, T., (2019). Thermochemical route for biohydrogen production. In: Pandey, A., (ed.), *Biohydrogen*. Elsevier, USA.

Kim, J. O., Kim, Y. H., Ryu, J. Y., Song, B. K., Kim, I. H., & Yeom, S. H., (2005). Immobilization methods for continuous hydrogen gas production biofilm formation versus granulation. *Process Biochem., 40*, 1331–1337.

Klass, D. L., (1998). Energy consumption, reserves, depletion and environmental issues. In: *Biomass for Renewable Energy, Fuels, and Chemicals* (pp. 1–27) Academic Press, San Diego, USA.

Kotay, S. M., & Das, D., (2008). Biohydrogen as a renewable energy resource—prospects and potentials. *Int. J. Hydrogen Energy, 33*, 258–263.

Kruse, A., (2009). Hydrothermal biomass gasification. *J. Supercrit. Fluids, 47*, 391–399.

Lan, P., Xu, Q., Zhou, M., Lan, L., Zhnag, S., & Yan, Y., (2010). Catalytic steam reforming of fast pyrolysis bio-oil in fixed bed and fluidized bed reactors. *Chem. Eng. Technol., 33*, 2021–2028.

Levin, D. B., Pitt, L., & Love, M., (2004). Biohydrogen production: Prospects and limitations to practical application. *Int. J. Hydrogen Energy, 29*, 173–185.

Li, X., Shen, Y., Wei, L., He, C., Lapkin, A. A., Lipinski, W., Dai, Y., & Wang, C. H., (2020). Hydrogen production of solar-driven steam gasification of sewage sludge in an indirectly irradiated fluidized-bed reactor. *Appl. Energy, 261*, 1–17.

Liu, Q., Hong, H., Yuan, J., Jin, H., & Cai, R., (2009). Experimental investigation of hydrogen production integrated methanol steam reforming with middle-temperature solar thermal energy. *Appl. Energy, 86*, 155–162.

Manish, S., & Banerjee, R., (2008). Comparison of biohydrogen production processes. *Int. J. Hydrogen Energy, 33*, 279–286.

Maschio, G., Lucchesi, A., & Stoppato, G., (1994). Production of syngas from biomass. *Bioresour. Technol., 48*, 119–126.

McKendry, P., (2002). Energy production from biomass (part 1): Overview of biomass. *Bioresour. Technol., 83*, 37–46.

Minowa, T., Zhen, F., & Ogi, T., (1998). Cellulose decomposition in hot-compressed water with alkali or nickel catalyst. *J. Supercrit. Fluids, 13*, 253–259.

Nanda, S., Dalai, A. K., Berruti, F., & Kozinski, J. A., (2016a). Biochar as an exceptional bioresource for energy, agronomy, carbon sequestration, activated carbon and specialty materials. *Waste Biomass Valor., 7*, 201–235.

Nanda, S., Rana, R., Zheng, Y., Kozinski, J. A., & Dalai, A. K., (2017). Insights on pathways for hydrogen generation from ethanol. *Sustain. Energy Fuels, 1*, 1232–1245.

Nanda, S., Reddy, S. N., Dalai, A. K., & Kozinski, J. A., (2016b). Subcritical and supercritical water gasification of lignocellulosic biomass impregnated with nickel nanocatalyst for hydrogen production. *Int. J. Hydrogen Energy, 41*, 4907–4921.

Nanda, S., Reddy, S. N., Hunter, H. N., Vo, D. V. N., Kozinski, J. A., & Gökalp, I., (2019). Catalytic subcritical and supercritical water gasification as a resource recovery approach from waste tires for hydrogen-rich syngas production. *J. Supercrit. Fluids, 154*, 104627.

Okolie, J. A., Nanda, S., Dalai, A. K., Berruti, F., & Kozinski, J. A., (2020). A review on subcritical and supercritical water gasification of biogenic, polymeric and petroleum wastes to hydrogen-rich synthesis gas. *Renew. Sust. Energy Rev., 119*, 109546.

Okolie, J. A., Rana, R., Nanda, S., Dalai, A. K., & Kozinski, J. A., (2019). Supercritical water gasification of biomass: A state-of-the-art review of process parameters, reaction mechanisms and catalysis. *Sustain. Energy Fuels, 3*, 578–598.

Osada, M., Sato, T., Watanabe, M., Adschiri, T., & Arai, K., (2004). Low-temperature catalytic gasification of lignin and cellulose with a ruthenium catalyst in supercritical water. *Energy Fuels, 18*, 327–333.

Pandu, K., & Joseph, S., (2012). Comparisons and limitations of biohydrogen production processes: A review. *Int. J. Adv. Eng. Technol., 2*, 342.

Parthasarathy, P., Narayanan, S., Ceylan, S., & Pambudi, N. A., (2017). Optimization of parameters on the generation of hydrogen in combined slow pyrolysis and steam gasification of biomass. *Energy Fuels, 31*, 13692–13704.

Peng, W. X., Wang, L. S., Mirzaee, M., Ahmadi, H., Esfahani, M. J., & Fremaux, S., (2017). Hydrogen and syngas production by catalytic biomass gasification. *Energy Convers. Manag., 135*, 270–273.

Pinto, C., Gulyurtlu, F., & Carbita, I., (2003). The study of reactions influencing the biomass steam gasification process. *Fuel, 82*, 835–842.

Pinto, T., Gouveia, L., Ortigueira, J., Saratale, G. D., & Moura, P., (2018). Enhancement of fermentative hydrogen production from *Spirogyra sp.* by increased carbohydrate accumulation and selection of the biomass pretreatment under a biorefinery model. *J. Biosci. Bioeng., 126*, 226–234.

Pio, D. T., Tarelho, L. A. C., & Matos, M. A. A., (2017). Characteristics of the gas produced during biomass direct gasification in an autothermal pilot-scale bubbling fluidized bed reactor. *Energy, 120*, 915–928.

Rana, R., Nanda, S., Maclennan, A., Hu, Y., Kozinski, J. A., & Dalai, A. K., (2019). Comparative evaluation for catalytic gasification of petroleum coke and asphaltene in subcritical and supercritical water. *J. Energ. Chem.*, 107–118.

Rasmussen, N. B. K., & Aryal, N., (2020). Syngas production using straw pellet gasification in fluidized bed allothermal reactor under different temperature conditions. *Fuel, 263*, 1–8.

Reddy, S. N., Nanda, S., Dalai, A. K., & Kozinski, J. A., (2014). Supercritical water gasification of biomass for hydrogen production. *Int. J. Hydrogen Energy, 39*, 6912–6926.

Salimi, M., Safari, F., Tavasoli, A., & Shakeri, A., (2016). Hydrothermal gasification of different agricultural wastes in supercritical water media for hydrogen production: A comparative study. *Int. J. Ind. Chem., 7*, 277–285.

Sarangi, P. K., & Nanda, S., (2020). Biohydrogen production through dark fermentation. *Chem. Eng. Technol., 43*, 601–612.

Shanmugam, S., Hari, A., Pandey, A., Mathimani, T., Felix, L., & Pugazhendhi, A., (2020). Comprehensive review on the application of inorganic and organic nanoparticles for enhancing biohydrogen production. *Fuel, 270*, 117453.

Show, K. Y., Lee, D. J., Tay, J. H., Lin, C. Y., & Chang, J. S., (2012). Biohydrogen production: Current perspectives and the way forward. *Int. J. Hydrogen Energy, 37*, 15616–15631.

Solunke, R. D., & Veser, G., (2010). Hydrogen production via chemical looping steam reforming in a periodically operated fixed-bed reactor. *Ind. Eng. Chem. Res., 49*, 11037–11044.

Sui, M., Li, G., Guan, Y., Li, C., Zhou, R., & Zarnegar, A., (2020). Hydrogen and syngas production from steam gasification of biomass using cement as catalyst. *Biomass Convers. Bioref., 10*, 119–124.

Sun, H., & Wu, C., (2019). Autothermal CaO looping biomass gasification for renewable syngas production. *Environ. Sci. Technol., 53*, 9298–9305.

Takeuchi, Y., Yamamoto, Y., & Konishi, S., (2007). Hydrogen production from biomass using high temperature nuclear heat. *Fusion. Sci. Technol., 52*, 756–760.

Udomchoke, T., Wongsakulphasatch, S., Kiatkittipong, W., Kiatkittipong, W., Arpornwichanop, A., Khaodee, W., Powell, J., et al., (2016). Performance evaluation of sorption enhanced chemical-looping reforming for hydrogen production from biomass with modification of catalyst and sorbent regeneration. *Chem. Eng. J., 303*, 338–347.

Wang, S., Zhang, F., Cai, Q., Li, X., Zhu, L., Wang, Q., & Luo, Z., (2014). Catalytic steam reforming of bio-oil model compounds for hydrogen production over coal ash supported Ni catalyst. *Int. J. Hydrogen Energy, 39*, 2018–2025.

Wu, H., Li, J., Liao, Q., Fu, Q., & Liu, Z., (2020). Enhanced biohydrogen and biomethane production from *Chlorella sp.* with hydrothermal treatment. *Energy Convers. Manage, 205*, 112373.

Xiao, Y., Xu, S., Song, Y., Shan, Y., Wang, C., & Wang, G., (2017). Biomass steam gasification for hydrogen-rich gas production in a decoupled dual loop gasification system. *Fuel. Process. Technol., 165*, 54–61.

Yao, S., Zhang, X., Zhou, W., Gao, R., Xu, W., Ye, Y., Lin, L., et al., (2017). Atomic-layered Au clusters on α-MoC as catalysts for the low-temperature water-gas shift reaction. *Science, 357*, 389–393.

Ye, T., Yuan, L., Chen, Y., Kan, T., Tu, J., Zhu, X., Torimoto, Y., et al., (2009). High efficient production of hydrogen from bio-oil using low-temperature electrochemical catalytic reforming approach over NiCuZn-Al_2O_3 catalyst. *Catal. Lett., 127*, 323–333.

Yuan, L., Chen, Y., Song, C., Ye, T., Guo, Q., Zhu, Q., Torimoto, Y., & Li, Q., (2008). Electrochemical catalytic reforming of oxygenated-organic compounds: A highly efficient method for production of hydrogen from bio-oil. *Chem. Commun., 41*, 5215–5217.

Zhang, Y., Zhang, T., Zhang, Z., Tahir, N., & Zhang, Q., (2020). Biohydrogen production from *Humulus scandens* by dark fermentation: Potential evaluation and process optimization. *Int. J. Hydrogen Energy, 45*, 3760–3768.

Zhu, J., Li, Y., Wu, X., Miller, C., Chen, P., & Ruan, R., (2009). Swine manure fermentation for hydrogen production. *Bioresour. Technol., 100*, 5472–5477.

CHAPTER 2

Production of Hydrogen through Gasification Technology

PARESH H. RANA

Department of Chemical Engineering, Lalbhai Dalpatbhai College of Engineering, Ahmedabad, Gujarat, India
E-mail: ranaph78@gmail.com

ABSTRACT

To overcome CO_2 emissions and environmental problems shaped due to extensive usage of fossil fuel-based feedstock, it is essential to produce potential chemicals through the cleanest energy carrier. The industrial importance of hydrogen is not needed to be emphasized. It can be obtained from biomass *via* thermochemical routes, which have the potential for commercial implementation as compared to other treatment routes of biomass. Biomass conversion to produce hydrogen *via* pyrolysis and different gasification technologies are mostly investigated. The catalytic steam reforming is attaining significant interest owing to its optimized control in catalytic and pyrolysis steps as compared to gasification. Reforming, sorption enhanced and chemical looping processes are used for removal of generated CO_2 (*in situ)* and to produce high purity H_2 gas. These processes are believed to shift the thermodynamic limits of WGS (water-gas shift) reaction, which enhances the conversion of feedstock and process effectiveness. The CO_2 capture and autothermal operating conditions can be attained through the cyclic redox mechanism of oxygen carriers (OCs) in hydrogen production by chemical looping processes. This chapter deals with biomass conversion to hydrogen through gasification technology. It also discusses the process parameters such as type of biomass feedstock types, particle size, moisture content, temperature, pressure, steam/biomass ratio, equivalence ratio and catalyst-to-sorbent ratio, which have a considerable impact on the product quality. The challenges and perspective of conversion of biomass to hydrogen are also described.

2.1 INTRODUCTION

Primary energy sources like oil, natural gas, and coal are being used to meet our energy requirements since the industrial revolution. However, the exhaustive usage of these energy sources not only depletes reserves but also creates environmental issues. Globally CO_2 emission is increasing dramatically through the consumption of oil, gas, and coal for combustion-related activities. To overcome these issues, in recent years, researchers have made constant efforts to alternate conventional resources (such as fossil fuels) with non-conventional (renewable energy) resources that are considered clean energies. Due to their capability to improve energy sustainability and reduce CO_2 emission, biomass, and hydrogen have received substantial gained considerable interest among the renewable resources (Ahmadi et al., 2013).

Biomass is considered the worlds' fourth largest energy source after oil, gas, and coal, and is estimated to be equivalent to about 14% of global primary energy (Farzad et al., 2016). With increased throughput, the development of technologies and low levels of CO_2 emission makes it an attractive alternative to conventional sources. It is an abundant and low-cost renewable material having substantial amounts of hydrogen, oxygen, and carbon (Okolie et al., 2021). It is obtained from a wide range of sources such as agricultural waste, wood, and wood wastes, crop residues, animal waste, municipal solid waste, oceanic plants and organic wastes (Salam et al., 2018).

Hydrogen (H_2) has the potential to be a clean alternative energy carrier to fossil fuels for various applications such as electricity production, fuel cells, transportation, etc. (Hossain and Charpentier, 2015; Nanda et al., 2017). For instance, in fuel cells to produce electricity. The hydrogen energy content is approximately 2.75 times that of hydrocarbon fuels and is around 122 kJ/g (Hosseini et al., 2015). H_2 fuel cell vehicles are 3 times more efficient than gasoline engines (Momirlan and Veziroglu, 2005). Apart from its capability as an energy carrier, H_2 is also used in petrochemical, electronics processing, food, and metallurgical industries as a chemical feedstock.

Currently, H_2 is derived from non-conventional fossil fuel feedstocks. From approximately 96% H_2 produced from conventional sources, nearly 30% and 48% are obtained from oil steam reforming and natural gas, respectively, whereas coal gasification with WGS reaction contributes to around 18% H_2. While, water electrolysis produced 4% H_2 (Arregi et al., 2018; Qiu et al., 2018). The production of H_2 from non-renewable resources is not environmentally benign due to the emission of excess CO_2. Therefore, to curb CO_2 emission, H_2 should be obtained from renewable resources that are believed to emit less amount of CO_2. Electrolysis, biomass, and solar conversion are renewable methods through which H_2 is produced. In the electrolysis process, H_2 and O_2 are generated from the splitting of H_2O using electrical energy from non-conventional sources. Biomass either thermo-chemically or biologically converted to H_2 rich gas, while H_2 is produced by thermal decomposition assisted by solar heat or photolysis using electro-chemical systems.

This chapter discusses several hydrogen production technologies. The production of H_2 through various gasification technologies and the effect of parameters, i.e., temperature, fuel/air ratio, pressure, biomass characteristics, catalysts, and sorbents on H_2 production are also discussed (Okolie et al., 2019). Figure 2.1 depicts various methods used for H_2 production from biomass-derived feedstocks. Biomass can be converted to hydrogen (bio-hydrogen) *via* thermochemical, biological, electrochemical, and photochemical processes. The thermochemical method for H_2 production is a simpler route as compared to chemical and biochemicals methods and not required the addition of chemicals. The production of H_2 greatly depends on the various parameters such as biomass feedstocks and their characteristics, operating conditions, catalysts, sorbents, etc. (Huda et al., 2014).

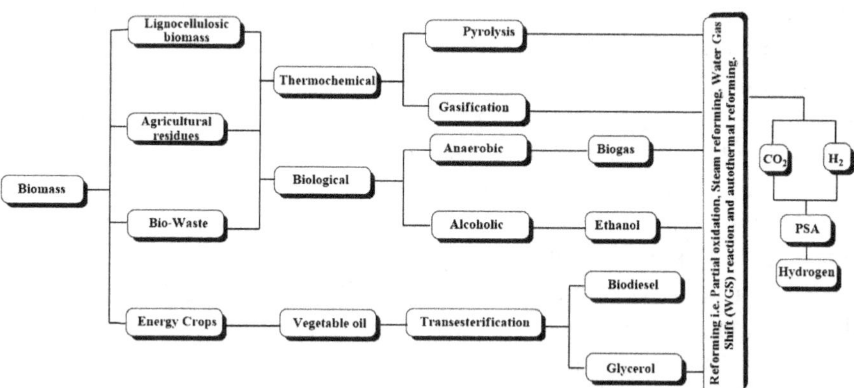

FIGURE 2.1 Various methods for the production of hydrogen production from biomass feedstocks.

Thermochemical processes mainly include gasification, pyrolysis, combustion, and liquefaction (Mohanty et al., 2015; Salam et al., 2018; Shayan et al., 2018; Nanda and Berruti, 2021). Among them, gasification technology is believed to be the most prominent path to produce H_2. However, the production cost of H_2 ($1.21–2.42 per kg) from biomass pyrolysis and gasification is reported very high than from methane steam reforming ($0.75 per kg, without CO_2 sequestration) (Parthasarathy and Narayanan, 2014).

2.2 PYROLYSIS

Biomass is thermally degraded to gases, char, and condensable products in pyrolysis without the use of oxygen. However, product proportions were found to be dependent on a broad range of process parameters, i.e., type of biomass, temperature, residence time, and particle size. Low temperatures, high vapor residence times and low heating rates favor charcoal production, while long residence times, high temperatures and slow heating rates increase the production of gas, and high heating rates, short residence time and moderate temperatures are optimum for liquid products (Cheng et al., 2012; Parthasarathy and Sheeba, 2015). The product property and effect of pyrolysis parameters, reaction mechanisms for lignocellulosic biomass pyrolysis have been studied and reviewed (Anca-Couce, 2016; Kan et al., 2016; Wang et al., 2017). However, H_2 concentration in the product gas is too low to be commercially attractive.

Catalytic pyrolysis of biomass is one of the alternatives to enhance the yield of H_2. Sun et al. (2016) employed ZSM-5 and Fe/ZSM-5 to produce hydrocarbons *via* fast pyrolysis of biomass and found that Fe/ZSM-5 exhibited higher catalytic activity than ZSM-5. To order to improve H_2 production, different char-based catalysts tested for pyrolysis of olive pomace using steam in a two-stage fixed bed reactor. The obtained results demonstrated that in the absence of a catalyst, steam does not influence H_2 production. It was also observed that increased bed temperature from 500°C to 700°C led to improved production of H_2 in the presence of different char-based catalysts (Duman and Yanik, 2017).

Lu et al. (2018) studied carbides and nitrides of tungsten and molybdenum supported on activated carbon for fast pyrolysis of pinewood to obtain high-grade oil. The obtained results showed that the catalytic performance of W_2C/AC and W_2N/AC catalysts were in line with Pd/AC catalyst, while Mo_2C/AC and Mo_2N/AC did not display at par catalytic activity. It

TABLE 2.1 Summary of the Notable Works on Pyrolysis and Gasification of Biomass

Process	Feedstock	Reaction Conditions	Reactor	Product Yield	References
Flash pyrolysis	Woody biomass	• Temperature: 400–500°C • Vapor residence time: 4 s	Flow reactor	Bio-oil and gases	Imran et al. (2016)
Fast pyrolysis	Algae (*Chlorella vulgaris*)	• Temperature: 700°C	Fixed bed reactor	• Gas: 46.8% • Bio-oil: 27.3% • Char: 23.7%	Yuan et al. (2015)
Slow pyrolysis	Pinewood chips	• Temperature: 700°C • Heating rate: 20°C/min	Gray King pyrolysis assay	• Gas: 42% • Bio-oil: 18% • Biochar: 42%	Russell et al. (2017)
Steam gasification	Wood residue	• Temperature: 700°C • Reaction time 40 min • Steam-to-biomass (S/B) ratio: 1 • Biomass particle size: 1–2.5 mm	Fluidized bed reactor	• H_2: 50% • CO: 20% • CO_2: 18–19% • CH_4: 10% • C_2, C_3, and C_{4+}: Traces	Fremaux et al. (2015)
Air gasification	Palm oil	• Temperature: 700–1000°C, • Atmospheric pressures • Biomass particle size: 0.3–1 mm	Fluidized bed reactor	High temperature favored H_2 and gas yields	Mohammed et al. (2011)
Steam-air gasification	Sesame wood	• Temperature: 700–900°C • Equivalence ratios: 0.18	Downdraft gasifier	H_2 and CO_2 yields increased with steam addition	Sharma and Sheth (2016)

TABLE 2.1 *(Continued)*

Process	Feedstock	Reaction Conditions	Reactor	Product Yield	References
Steam-oxygen gasification	Casuarina wood	• Temperature: 727°C • S/B ratio: 1 • Equivalence ratio: 0.05	Open-top downdraft gasifier	• H_2: 54% • CO: 28% • CO_2: 16%; CH_4: 2%	Sandeep and Dasappa (2014)
Supercritical water gasification	Glucose	• Temperature: 400°C • Pressure: 24.5 MPa • Reaction time: 20 min • Catalysts: Ni/γAl_2O_3 and Ni/CeO_2-γAl_2O_3	Autoclave	Yield and selectivity of H_2 increased with catalyst	Lu et al. (2010)
Chemical looping gasification	Biomass char	• Temperature: 850°C	Fixed bed reactor	• H_2: 22.3% • CO: 44.6%; CO_2: 33.1%	Huang et al. (2016)
Solar gasification	Pine and spruce wood	• Temperature: 1000–1400°C	Tubular solar reactor	• H_2: 35–45% • Carbon conversion rate: 93.5%	Bellouard et al. (2017b)

was observed that lignin decomposed to phenolic compounds, which further catalyze by non-stable functional groups to produce stable phenolics in the presence of W_2C/AC and W_2N/AC catalyst. However, to maximize H_2 production in pyrolysis, preference should be given long residence time and high temperature along with catalyst employed. Some notable studies on pyrolysis and gasification of different feedstocks at varying reaction conditions and reactors are summarized in Table 2.1.

2.3 GASIFICATION

The conventional gasification process occurs at high temperatures (typically >700°C), which leads to produces gaseous products and charcoal. The gaseous products are mainly composed of H_2, CH_4, CO, CO_2, and other hydrocarbons. The charcoal is converted to H_2, CH_4, CO, and CO_2. Unlike pyrolysis, the gasification process occurs with gasifying agents such as steam, air, oxygen or a mixture of these components. Produced gases can be reformed by steam to enhance H_2 formation and later on reaction between CO and H_2O known as water-gas shift (WGS) reaction further increase H_2 production (Demirbaş, 2002; Ahmad et al., 2016). Series of complex thermochemical reactions are involved in this process. Therefore, it is not possible to separate the reactor into various zones to perform gasification reactions concurrently. As seen in Figure 2.2, drying, pyrolysis, combustion, reduction, and char gasification steps are involved in the gasification process.

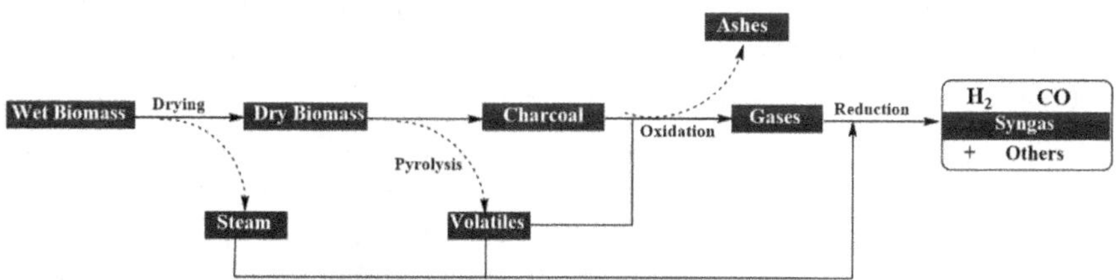

FIGURE 2.2 A typical biomass steam gasification process.

In the first step of biomass gasification, moisture from biomass feedstocks is removed with the help of drying at low temperatures. Thermal decomposition of biomass yields char and volatiles (gases and liquids) during pyrolysis. The pyrolysis liquid from the primary decomposition is a mixture of liquid organic components, which comprises alcohols, furans, phenols, acids, esters, ketones, aldehyde, etc., (Diblasi, 2008; Kan et al., 2016). Under gasification conditions, these unstable compounds are thermally or catalytically cracked to stable aromatic structures, i.e., benzene, toluene, and naphthalene generally considered as secondary and/or tertiary tar (Devi et al., 2005; Palma, 2013). Tar present in syngas causes operational problems by the formation of aerosols and polymerization into a complex structure. Besides, condensed tar leads to block particle filters and fuel lines, fouling, and corrosion that creates a problem in syngas valorization (Peng et al., 2017). Hence, the elimination of tar from syngas is essential for overall process performance and reactor design (Sansaniwal et al., 2017b). To obtain high tar cracking performance, a temperature of 1250°Cand 0.5 s of residence times were required (Palma, 2013).

Carbon is the main constituent of char (Nanda et al., 2016). Char gasification mainly occurs through water-gas shift reaction (reaction between CO and H_2O) and Boudouard reaction (reaction between carbon and CO_2). The reaction rate for both reactions is found to be low as compared to pyrolysis and is considered as a controlling step in the overall gasification process. Hence, to achieve high conversion an adequate reactor with suitable residence time is required (Ahmed and Gupta, 2011; Lopez et al., 2016). Therefore, to attain a high yield of H_2, it is essential to have high H_2 and low tar content from gasifier gas. Various reaction parameters like biomass characteristics, pressure, temperature, residence time, equivalence ratio and gasifier design can affect the quality of gas (Okolie et al., 2019).

Being an endothermic process, heat is supplied into the gasification process to maintain the system temperature. Heat can be provided either directly (autothermal) or indirectly (allothermal). Heat is produced directly through partial oxidation within the reactor in auto thermal gasification, whereas in all thermal gasification, heat is supplied to the gasifier from outside. The heat is generated through combustion and transfer with a heat exchanger or heat carrier, i.e., bed material (Abuadala and Dincer, 2012).

Kihedu et al. (2016) performed pelletized biomass gasification with auto-thermal updraft gasification using air and mixture of the air-stream at a 2.85 equivalence ratio using a packed bed reactor. Syngas comprising CO (26.8%), CO_2 (5.3%), H_2 (5%) and CH_4 (1.4%) was obtained with a heating value (HV) of 4.4 MJ/Nm^3 in air gasification. However, a marginal improvement in the heating value of syngas was observed with air-stream gasification and it was about 4.5 MJ/Nm^3. In general, low H_2 concentration and low HV syngas produce from biomass air gasification.

Biomass steam gasification (35 wt.% moisture content) with calcined dolomite, calcined limestone and a mixture of dolomite and limestone was applied using a fixed bed reactor in temperature from 450°C to 950°C. The results showed that the highest H_2 yield (204.6 mL/$g_{biomass}$) was obtained at reaction conditions (i.e., temperature: 850°C, dolomite/limestone mass ratio: 1, steam/biomass ratio: 0.5 and sorbent/biomass ratio: 0.7) (Zhang et al., 2015). The yield of H_2 and energy content are found better in steam gasification as compared to air gasification and fast pyrolysis (Balat, 2008; Parthasarathy and Narayanan, 2014).

Biomass oxygen gasification is another efficient method to produce H_2 rich gases with medium heating value. Wang et al. (2015) employed a two-stage gasification gasifier to produce clean syngas from biomass. The results showed that increased O_2 level has a positive impact on syngas production, but modest on the ratio of H_2/CO. The H_2 concentration and CO in syngas enhanced to 70 vol% from 30 vol% when the O_2 level changed 99.5 vol% from 21 vol%. However, the high investment cost for O_2 production is the major drawback of the process.

An experimental investigation was carried out with almond shells in a bubbling fluidized bed reactor (FBR) with steam/oxygen as a gasifying agent (Barisano et al., 2016). The results demonstrated that FBR with internal recirculation was able to produce H_2 (30–33 vol% dry gas) with 10.9–11.7 MJ/Nm^3 of LHV at 820–830°C and 1–1.1 bar pressure higher than air (35 and 50 wt.% O_2) enriched gasification (Barisano et al., 2016). This enhancement in product quality was not only due to the reduced amount of N_2 gas, but also due to a positive impact of steam addition. The use of steam fosters the WGS reaction of the gasification process to augment H_2 production. The requirement of steam temperature (>700°C) further increased the cost of steam generators.

Substantial technological development has taken place in biomass gasification technology (Sansaniwal et al., 2017a). Different types of gasifiers such as entrained flow, fixed bed (e.g., cross draft, downdraft, and updraft), spouted beds, fluidized bed (circulating and bubbling bed), plasma reactors and rotary kiln have been used in the gasification process (Heidenreich and Foscolo, 2015; Mahinpey and Gomez, 2016; Molino et al., 2016; Sikarwar et al., 2016; Bellouard et al., 2017a). A straightforward comparison between reactors configurations cannot be done due to the differences in the operating conditions (i.e., temperature, pressure, reaction time, feeding rate, biomass particle size, non-uniform biomass properties, and procedures for tar and gas analyzes). For example, H_2 production in 1.3–8.5 wt.% range along with tar contents (4–140 g/Nm^3) are reported with the employment of updraft fixed bed (Umeki et al., 2010), entrained flow (Hernández et al., 2012), rotary kiln (Iovane et al., 2013), free fall (Wei et al., 2007), spouted bed (Artetxe et al., 2010; Bellouard et al., 2017a) and fluidized bed (Gil and Corella, 1999; Göransson et al., 2011) reactors in non-catalytic biomass gasification. Low tar concentrations in the range of 20–80 g/Nm^3 along with high H_2 production and concentration of 6.9 wt.% and >50 vol%, respectively are reported (Gil and Corella, 1999; Göransson et al., 2011).

Fuel gas with a heating value of 4–7 MJ/Nm^3 was obtained using air from autothermal gasification while pure O_2 gasification produced syngas of 10–12 MJ/Nm^3 heating value. However, allothermal steam gasification in a dual fluidized bed reactor (DFBR) achieved a gas product with a heating value of 12–20 MJ/Nm^3 (Göransson et al., 2011). Herein, steam was blown into the gasifier and char was burnt in the combustor, to obtain high H_2 content syngas, and to heat the system, respectively (Göransson et al., 2011; Wilk and Hofbauer, 2013). This DFBR technology of two circulating and/or bubbling fluidized beds, where one is acting as a char combustor and another as a gasifier has satisfactorily solved the heat supply issue (Göransson et al., 2011; Yao et al., 2017). As DFB gasifier operated at >800°C and heat supplied by char combustion may not be enough, thus external supply for energy is needed to meet heat specifications (Corella et al., 2007).

2.4 EFFECTS OF VARIOUS PARAMETERS IN BIOMASS GASIFICATION

Various parameters such as particle size, moisture content, density, feedstock types, temperature, pressure, S/B ratio, sorbent/biomass ratio, equivalence ratio, catalyst, sorbents significantly affect the quality of the product obtained during biomass gasification.

2.4.1 CHARACTERISTICS OF BIOMASS

The biomass type, moisture content and particle size are the main characteristics affecting gasification performance. Biomass is majorly composed of cellulose, hemicelluloses, and lignin in the range of 30–60%, 20–35%, and 15–30%, respectively. Lignocellulose is an important material of biomass and is inexpensive, renewable, and abundant. It plays a crucial role in the thermo-chemical process for biomass valorization (Van de Velden et al., 2010). Generally, the ratio of cellulose/lignin and hemicellulose/lignin in biomass material is in the range of 0.5–2.7 and 0.5–2.0, respectively (Basu,

2010a). Burhenne et al. (2013) used rape straw, wheat straw and spruce wood as feedstock with a higher content of lignin, cellulose, and hemicellulose. It was observed that high cellulosic and hemicellulosic biomass material produced gaseous products, whereas oil was a major product in high lignin material.

In most cases, gasifiers are designed for moisture content (10–15 wt.%), although up to 35 wt.% moisture contents can be easily accommodated. However, high moisture contents of material reduce the operating temperature of the gasifier, which in turn reduces H_2 production and increases tar content in the product (Kaushal and Tyagi, 2012). Fluidized bed (bubbling and circulating) gasifier can work efficiently with feedstock contained moisture (<55%). The upward, downward, and entrained flow fixed bed gasifier can tolerate 50%, 20% and 15% maximum moisture content in the feedstock, respectively (Basu, 2010b; Arena, 2012).

The particle size of biomass has a considerable impact on H_2 production. The smaller particle size of biomass endows with higher surface area enhances mass and heat transfer between particles. This ultimately led to improve the rate of gasification reactions (carbon conversion reactions, water-gas shift reaction, and Boudouard reaction) and increased rate, leading to high H_2 concentration with gas yields and low tar contents (Kaushal and Tyagi, 2012; Parthasarathy and Sheeba, 2015).

Feng et al. (2011) checked the effects of particle size and operating temperature between 700°C and 900°C on product composition and yield using pine sawdust as biomass. They reported that an increase in temperature to 900°C from 800°C led to decomposed tar completely and the gas production rate increased with smaller size particles at a higher temperature (Feng et al., 2011). However, when particle size rose to 0.25 mm from 0.125 mm, H_2 concentration reduced from 58 vol% to 40 vol% in product gas at 900°C. A similar observation of increased yield of H_2 and dry gas, and carbon conversion efficiency and reduced tar and char content with lower particle size made by Luo et al. (2009) and Hernández et al. (2010). However, a reduction in particle size <1 mm requires a higher amount of energy (de Lasa et al., 2011; Kaushal and Tyagi, 2012).

2.4.2 TEMPERATURE

H_2 production is mostly influenced by temperature in the biomass steam gasification. The studies of temperature influence on compositions and yield carried out by various researchers confirmed that high gas yield and minimum tar were obtained with a high temperature of gasifier (Li et al., 2009; Skoulou et al., 2009; Erkiaga et al., 2014; Wang et al., 2017; Zhang and Pang, 2017). The higher gasifier temperature increased heating rates between particles which effectively led to decomposed particles and enhanced rates of reaction involved in gasification (e.g., steam reforming, Boudouard pyrolysis, water-gas shift and char steam gasification reactions) (Kaushal and Tyagi, 2012; Parthasarathy and Narayanan, 2014). This led to increasing content of H_2, CH_4, and CO in the product gas, which ultimately enhances the overall yield of the process.

Li et al. (2009) examined the influence of temperature between 750–900°C and particle size in the range of 0.15 mm and 5 mm using palm oil waste with a nano-NiLaFe/γ-Al_2O_3 catalyst in a fixed bed gasifier at atmospheric pressure. The obtained results showed that when the temperature rose to 900°C from 750°C, the yield of gas reached 2.5 $m^3/kg_{biomass}$ from 1.5 $m^3/kg_{biomass}$ and that of H_2 increased to 133.3 g $H_2/kg_{biomass}$ from 77.5 g $H_2/kg_{biomass}$. With temperature rise, the content of H_2 and CO_2 improved and that of CH_4 and CO declined.

Erkiaga et al. (2014) performed gasification of pinewood sawdust and evaluated the effects of S/B ratio (0–2), particle size (0.3–4 mm) and temperature (800–900°C) with spouted bed reactor. They observed a rise in temperature from 800°C to 900°C, the increased gas yield from 0.73 to 0.96 $Nm^3/kg_{biomass}$ and the concentration of H_2 enhanced from 28 vol% to 38 vol%. The improvement of reforming reactions at high temperatures led to a reduced concentration of CH_4 and hydrocarbons, which is in line with the observation made by Li et al. (2009). Higher reactor temperature (850°C) favored maximum H_2 production even in the non-catalytic study for pinewood and legume straw steam gasification. At 850°C, reductions in tar content were observed from 45.6 g/Nm^3 to 6 g/Nm^3 and from 62.8 g/Nm^3 to 3.7 g/Nm^3 for pinewood and legume straw, respectively (Wei et al., 2007).

Greater yields of gaseous products and H_2 concentration at elevated temperatures can be due to a higher rate of tar degradation. The tar and char were effectively decomposed at increased temperature by thermal cracking and Boudouard reactions (He et al., 2009). Hence, maintaining a high temperature in the gasifier can contribute effectively to biomass gasification when product gas is desired. Fixed bed gasifiers typically operate at 1100°C, whereas an upward draft gasifier requires a longer time as compared to a downward draft gasifier to reach the working temperature. A fluidized bed reactor is operated at temperatures below 1000°C to evade agglomeration and ash fusion while temperatures more than 1900°C are required for an entrained flow gasifier (Basu, 2010b).

2.4.3 *PRESSURE*

The operating pressure effect on product distribution during biomass gasification was investigated by several researchers(Knight, 2000; Mayerhofer et al., 2012; Radwan, 2012; Liu et al., 2016). Liu et al. (2016) performed gasification of rice husk at 700°C and 0.1–5 MPa pressure using FBR. They reported that the contents of CO and H_2 in gaseous products reduce while that of CO_2 and CH_4 improves with increased pressure (Liu et al., 2016).

Mayerhofer et al. (2012) investigated the influence of operating conditions such as temperature (750–840°C), S/B ratio (0.8–1.2) and pressure (0.1–0.25) on the quality of tar and gas in allothermal steam gasification in a bubbling FBR. They reported that a maximum of 5.7 g/m^3 of tar concentration was obtained with 0.25 MPa at S/B ratio of 1.2 and 750°C (Mayerhofer et al., 2012). Similar observations of the direct influence of pressure on the performance of the gasification process were made by Radwan (2012) and Newalkar et al. (2014). Fermoso et al. (2009) observed that the reaction rate of char gasification augmented linearly with the partial pressure of CO_2. At 1 bar and 10 bar pressure of CO_2 (partial), the order of reaction was found around 0.33 and 0.35, respectively. In contrast, Knight (2000) and Wolfesberger et al. (2009) reported that tar content was reduced by 20% and 30%, respectively with an increase in pressure at 825°C.

2.4.4 *STEAM-TO-BIOMASS RATIO*

Steam-to-biomass (S/B) ratio is another essential factor that strongly influences energy input, product yields, and outlet gas quality. Amounts of CH_4 and char elevated at a lower S/B ratio, while H_2 and CO_2 are major products at a high S/B ratio. Higher S/B ratio enhanced rates of reforming and WGS reactions and reduced CO content in product gas (Parthasarathy and Narayanan, 2014; Fremaux et al., 2015). Umeki et al. (2010) tested woody biomass with steam gasification at high temperatures using a fixed bed gasifier (updraft) with a 1.2 t/day feed rate. They concluded that at 2.15 S/B ratios, a maximum of 6.4 wt.% of H_2 was obtained and an increase in tar content from 50 g/Nm^3 to 100 g/Nm^3 was observed with increased S/B ratio from 1.4 to 2.7, respectively.

Iovane et al. (2013) carried out palm shells steam gasification in a rotary kiln with different S/B ratios in the range of 0.6–1 at 850°C. They concluded that H_2 composition in products augmented to 52% from 40% and CH_4 content reduced from 12% to 6% with increased S/B ratio. The presence of steam conventionally performs methane steam reforming reaction for H_2 production.

$$CH_4 + H_2O \longrightarrow CO + 3H_2 \quad (2.1)$$

Göransson et al. (2011) performed allothermal gasification with FBR with different S/B ratios (0.3, 0.6, and 0.9) at temperatures (750–850°C). The results demonstrated that H_2 yield increased and that of CH_4 and CO decreased with increased S/B at 750–850°C temperature range. Thus, an H_2 concentration of 51 vol% and tar contents of 10 g/Nm^3 were obtained at selected operating conditions (Göransson et al., 2011).

A positive impact of the S/B ratio on the composition of tar and H_2 was reported by Gil et al. (1999). However, Li et al. (2009) concluded that H_2 yield enhanced to 133.3% from 50.2% to when the S/B ratio was raised to 1.33 from 0. Further enrichment in the S/B ratio to 2.7 reduced the yield of H_2 as the presence of excess steam lowering reaction temperature degraded the quality of the gas production and increased the overall energy consumption of the process. Therefore, it is essential to find an optimal value of the S/B ratio (Li et al., 2009; Kaushal and Tyagi, 2012; Parthasarathy and Narayanan, 2014). A similar observation of an increase in the ratio of S/B, beyond threshold limit reduced process efficiency was made by Sharma and Sheth (2016).

2.4.5 EQUIVALENCE RATIO

Equivalence ratio (ER) is another vital parameter of the gasification process and is actual air used divided by theoretical air required for complete combustion of biomass. ER has a strong influence on gasification products. High ER results, high CO_2 content and a reduced fraction of H_2 and CO in the gaseous product. However, an increased ER value results in a lower heating value of syngas. However, a high ER leads to a reduction in tar concentration owing to the higher availability of O_2 that reacts with volatiles (Hanping et al., 2008; Palancar et al., 2009; Mohammed et al., 2012). Zhou et al. (2000) reported a minor increase in NH_3 yield upon increased ER from 0.25 to 0.37

at 800°C using sawdust as feedstock. They also reported that the ER did not strongly influence nitrogen species.

The ER value near 0.26 enhances efficiency in FBR due to high combustion heat at higher temperatures. In general, the desired optimum value is 0.2–0.3 for ER. It was observed that lower ER value (<0.2) resulted in insufficient combustion of biomass which led to forming more char with a low heating value of gas, whereas higher attained complete combustion at reduced efficiency (Wang et al., 2015; Hamad et al., 2016; Din and Zainal, 2016).

2.4.6 CATALYSTS

The industrial importance of catalysts needs not to be emphasized. It reduces thermal and mass transfer resistance through particles which in turn improves the performance of gasification reactions (i.e., Boudouard, combustion, WGS, CH_4 steam reforming) in biomass gasification. Increased efficiency of reaction (i.e., CO shift, WGS, and steam methane reforming) contribute to enhancing the yield of CO and H_2. Apart from this, catalysts also help in the elimination of tar in the product. Catalysts can work *in situ* and post-gasification reactions in biomass gasification (Dayton, 2002; Basu, 2010c; Parthasarathy and Narayanan, 2014). Catalysts used in this can be classified as primary and secondary catalysts. Catalyst, which is directly added to the biomass within the gasifier, is known as the primary catalyst, whereas secondary catalyst is performed from outside the gasifier (secondary reactor). The secondary reactor is located downstream from the pyrolysis gasifier (Ates et al., 2006; de Andrés et al., 2011; Sutton et al., 2001). Commonly, (i) natural catalysts such as olivine, dolomite, and CaO; (ii) alkali catalysts, i.e., Na, Li, K, Cs, and Rb; and (iii) metal catalyst such as Ni are the main three group of catalysts employed in biomass gasification of removal of tar from syngas (Sutton et al., 2001; Wang et al., 2008; Weerachanchai et al., 2009).

Dolomite as a primary catalyst has attracted much attention due to its availability and activity (Devi et al., 2005). However, it can be used for tar reduction in the gaseous product as a secondary catalyst (Xu et al., 2010). For tar reduction, calcined dolomite was found to be more active than non-calcined due to the presence of oxides on the surface and higher surface area (Xu et al., 2010). Hu et al. (2006) carried out apricot stone steam gasification at 800°C and atmospheric pressure with an S/B ratio of 1 using a fixed bed reactor and compared dolomite activity, calcined at 900°C with an uncalcined. The results demonstrated that calcined dolomite was found more efficient than uncalcined for increasing H_2 yield in the product gas.

Wei et al. (2007) tested dolomite, limestone, and olivine catalyst in a solid-gas concurrent downflow free-fall reactor. Among the used catalyst, dolomite exhibited high H_2 composition and lower tar in the product gas. However, the softness and frailty of dolomite erode it very easily that prevents its utilization, especially in FBR. Therefore, it is essential to use catalysts with high attrition resistibility (Dayton, 2002; Min et al., 2011; Wang et al., 2008).

Olivine is another natural catalyst that is successfully employed in a fluidized bed gasifier as a primary catalyst for tar decomposition owing to its high mechanical strength and tar reduction ability (Dayton, 2002; Devi et al., 2005). Koppatz et al. (2011) compared the activity of olivine and silica sand as the primary catalyst for steam biomass gasification with DFBR and found that olivine as compared to silica sand not only exhibited enhanced yield of H_2 and gas by favoring CO shift reaction but also reduced tar content by 35% (Koppatz et al., 2011).

Erkiaga et al. (2013) studied γ-alumina and olivine catalyst for tar removal in continuous steam gasification of pinewood using a conical spouted bed reactor and observed that tar content was reduced by 84% and 79% with γ-alumina and olivine, respectively. They also observed that both catalysts promote CO shift, tar reforming and cracking reaction, which ultimately led to enhanced content of H_2 and CO_2 (Erkiaga et al., 2013).

The addition effect of Ni (3.9 wt.%) (Michel et al., 2011) and Fe (10 wt.%) in tar cracking activity of olivine catalyst in an FBR was studied (Rapagnà et al., 2011). It was observed that the presence of Ni to olivine, improved gas content to 51% from 38% (without Ni), increased dry gas yield to 1.7 m^3/kg from 1 m^3/kg at 800°C and also effectively removed tar (Michel et al., 2011). With Fe/olivine catalyst, an overall improvement in gasification performance was observed. Fe/olivine exhibited 53% rise in H_2 concentration with a 40% increase in gas yield, also reduced CH_4 content by 24% and tar reforming extended by 61% on average as compared to bare olivine (Rapagnà et al., 2011). Nickel-based catalysts are widely used in industries, have demonstrated excellent catalytic activity in the decomposition of NH_3 in coal, gasification of biomass and tar destruction. They can eliminate tar effectively as compared to the natural mineral catalyst as a secondary catalyst (Xu et al., 2010).

Degado et al. (1996) tested commercial $Ni/MgAl_2O_4$ catalysts and calcined limestone, magnesites, and dolomites as a secondary catalyst in a two-stage gasifier, i.e., a primary (fluidized bed) reactor and a secondary (fixed bed) reactor. They observed that Ni/Mg-Al_2O_4 catalysts performed better than other used catalysts for the removal of tar, gas yield, and H_2/CO ratio. With Ni/MgAl_2O_4 catalysts about 2.3 $Nm^3/kg_{biomass}$ gas yield, 2.6% H_2/CO and < 0.5 g/Nm^3 tar content were obtained in exit gas (Delgado et al., 1996).

A similar observation of the high activity of Ni-based catalysts than dolomite was reported by Rapagná et al. (1998) using an identical reactor arrangement. The authors observed a small decreased in catalytic activity due to coke deposition after 2 h of operation. Arregi et al. (2018) reviewed the performance of majorly Al_2O_3 supported and alkali promoted (K, Ca, Mg or La) various Ni-based catalysts for H_2 production in steam gasification of biomass.

Sutton et al. (2001) used 5 wt.% Ni-supported on Al_2O_3, ZrO_2, TiO_2, SiO_2 and MOR1 catalysts synthesized by wet impregnation and two catalysts with different Ni:Al ratio by co-precipitation method and compared their activity for tar removal from peat under N_2 environment. Ni/TiO_2 and Ni/ZrO_2 catalysts were able to attain 98.1% and 95.2% tar conversion, respectively, while the co-precipitated catalyst gave tar conversion about 92% at 800°C. However, alumina-supported catalysts were capable of removing CH_4 from the exit gas stream, thus yielding a higher H_2/CO ratio compared to other used catalysts.

Srinakruang et al. (2005) studied the effectiveness of Ni-supported on Al_2O_3, a mixed oxide of SiO_2-Al_2O_3 and dolomite in suppressing carbon deposition in gasification. The obtained results showed demonstrated that 15 wt.% Ni/Al_2O_3 displayed higher catalytic activity than Ni/dolomite and Ni/SiO_2–Al_2O_3 catalyst in toluene steam reforming. On the other side, Ni/dolomite exhibited high resistance against coking that was due to dolomite basicity (Srinakruang et al., 2005).

Pfeifer and Hofbauer (2008) tested three K_2O or MgO-doped NiO/Al_2O_3 (16–40 wt.%) catalysts in DFBR followed by a secondary downstream reactor (i.e., fixed bed). The results demonstrated a more than 99% decrease in tar content, an average 45 vol% of H_2 composition in the product gas and LHV of 12 MJ/Nm^3 over 12 h of reaction time with 40% NiO/Al_2O_3 at 900°C. A decreased in coke formation by 25% was observed with increased temperature from 850°C to 900°C (Pfeifer and Hofbauer, 2008).

Peng et al. (2017) prepared Ni/CeO_2/Al_2O_3 and Ni/Al_2O_3 catalysts with different Ni loadings and investigated them in an FBR in wood residue steam gasification. They found that Ni/CeO_2/Al_2O_3 catalyst was exhibited increased syngas yield and decreased tar and char content than Ni/Al_2O_3 catalyst (Peng et al., 2017). Although the Ni-based catalysts display a good performance on tar reforming, relatively expensive supports and energy-intensive processes are used in their preparation. Char is a low-cost adsorbent that can be effectively used as a catalyst for tar elimination (Shen et al., 2015; Wang et al., 2011).

Wang et al. (2011) examined char (wood and coal), and Ni/char as a catalyst in sawdust gasification with updraft biomass gasifier in syngas production and tar elimination. Ni-supported catalysts were prepared by mixing char particles and NiO mechanically. It was observed that with Ni (15% NiO) based catalyst greater than 97% tar was removed from syngas with a residence time of 0.3 s at 800°C.

Shen et al. (2015) performed tar conversion by co-pyrolysis using Ni-Fe catalyst supported on RHA (rice husk ash) and rice husk char (RHC) at 800°C. They observed that under optimized conditions, 92.3% tar conversion efficiency was achieved by RHC Ni-Fe catalyst. However, the monometallic RHA Ni catalyst exhibited about 93% tar conversion. This revealed that Ni-supported catalysts possessed high tar decomposition efficiency. The efficient removal of tar is considered one of the major concerns in the valorization of biomass. The presence of tar creates operational problems, catalyst deactivation and produces carcinogenic compounds. Hence, to increased H_2 content and remove tar in syngas alkali metal catalysts widely studied in the gasification process of biomass.

Huang et al. (2012) employed Fe/CaO catalysts for H_2 production in continuous feeding fluidized bed gasifier. They observed that metal support interactions of Fe and CaO promote carbon conversion to 97% and increased the promotion ratio of H_2 to 70.4% approximately and prevent CaO deactivation by tar. Xu et al. (2018) investigated Fe/CaO with mass ratios of 5%, 10%, 15%, and 20% in an FBR for increasing H_2 content and yield of syngas. Among the used catalysts, 5% Fe/CaO was found most efficient in biomass gasification with 38.2 $mol/kg_{biomass}$, 26.4 $mol/kg_{biomass}$, 8.69 MJ/kg and 49.3% of syngas yield, H_2 yield, LHV values and gasification efficiency, respectively.

2.5 SUPERCRITICAL WATER GASIFICATION

Supercritical water gasification (SCWG) is a hydrothermal method of H_2 production for high moisture (>30 wt.%) content biomass like algae, agricultural wastes, black liquor, switchgrass, industrial effluents, manure, sewage sludge, etc., without feedstock drying (Guo et al., 2010; Sikarwar et al., 2016). The fluid phase of water beyond its critical points is termed supercritical water. The gaseous product of SCWG of biomass chiefly consists of H_2, CO, CH_4 and CO_2. The formed CO further converts into CO_2 via WGS reaction and methanation reaction (Guo et al., 2010; Yanik et al., 2007). It was observed that at higher temperatures (>600°C), H_2 was a major product, whereas at lower temperatures (<450°C), CH_4 was a major product(Feng et al., 2004; Guo et al., 2010).

D'Jesús et al. (2006) achieved nearly 100% conversion of clover grass and corn silage in SCWG at a temperature > 700°C and showed that increased temperature had a positive impact on H_2 production. Chen et al. (2013) studied SCWG of sewage sludge with different alkali catalysts (e.g., NaOH, KOH, K_2CO_3 and Na_2CO_3) in FBR. The results indicated that the alkali catalyst improved carbon gasification efficiency. At 540°C and pressure of 25 MPa, maximum molar fraction and H_2 yield of 56% and 15.5 mol/kg, respectively, were achieved with KOH. Thus, appropriate catalysts can lower reaction temperature and pressure, thereby minimizing the operating and fixed cost as well as increase H_2 yields and carbon conversion (Azadi and Farnood, 2011).

Mehrani et al. (2015) conducted SCWG of bagasse with Cu-promoted Ni/γ-Al_2O_3 catalyst at 400°C and 240 bar. The obtained results showed that upon increasing the loading of Cu and Ni up to 5 wt.% and 20 wt.%, respectively, H_2 in product gas improved to 58.1 mol% and the gas yield reached 20.8%. Hossain et al. (2017) carried out SCWG of glucose using Ru-Ni/Al_2O_3 catalyst at a temperature range of 400–500°C and pressure range of 25–35 MPa for H_2 production in a 600 cm^3 batch reactor. An overall enhancement in carbon gasification efficiency, H_2 production and destruction of total organic carbon with reduced coke generation attained with Ru-Ni/Al_2O_3 catalyst. Aerogel Ru-Ni/Al_2O_3 exhibited a higher H_2 yield (1.7 times) than xerogel Ru-Ni/Al_2O_3. Herein, the sol-gel synthesis method was used to prepare the aerogel Ru-Ni/Al_2O_3, whereas the xerogel catalyst was synthesized by the wetness impregnation method.

Elif and Nezihe (2016) carried out SCWG of fruit pulp in a batch reactor with Ru/C catalyst between 400°C to 600°C at 25 MPa. The results demonstrated that H_2 content in the product augmented to 54.1 from 12.2 mol/$kg_{biomass}$ when the temperature increased to 600°C from 400°C. Near-complete gasification of sugarcane bagasse to CH_4, H_2 and CO_2 was attained at 400°C with Ru supported on carbon and TiO_2 catalyst and obtained a higher yield of CH_4 and CO_2 (Osada et al., 2012).

Yakaboylu et al. (2018) attempted to industrialize SCWG of starch in an FBR of 50 kg/h capacity and results revealed that gasification efficiency and carbon gasification efficiency enhanced with increased temperature. The highest 73.9% carbon gasification efficiency was obtained with a 24.5 kg/h feed rate at 600°C. Although the catalysts efficiently increase gasification efficiency, reduce activation energy and enhance H_2 formation, problems like fouling, instability, poisoning, and decaying persist(Guo et al., 2010; Azadi and Farnood, 2011). High energy consumption and processing cost are required to maintain high temperatures and high pressure are major problems for the commercialization of the process (Correa and Kruse, 2018).

2.6 CHEMICAL LOOPING GASIFICATION

Chemical looping gasification (CLG) is an innovative low-cost technology to produce high-quality syngas from biomass feedstock. In the CLG process, the gasification agent is replaced with an oxygen carrier (OC). The system is consists of a fuel and air reactor. Herein, high-quality syngas is produce using lattice oxygen from the OC in the fuel reactor, and reduced OC regenerated in the air reactor. (Huang et al., 2013; Ge et al., 2016a;).

Haung et al. (2016) used steam and natural iron ore as gasification agents and oxygen carriers, respectively, in a fixed-bed gasifier for CLG of biomass char; concluded that deep reduction of oxygen carrier was avoided by adding steam. They obtained 2.5 g/L gas yield, 56.9% H_2 content and 64.2% carbon conversion with Fe_2O_3/H_2O. Huang et al. (2014b) performed CLG of biomass using natural iron as the OC in a thermogravimetric analyzer combined with bubbling FBR under an inert atmosphere. They observed that natural ore not only supplied lattice oxygen, but its strong redox reaction with pyrolysis products elevated the OC conversion. High reaction temperature enhanced the gas yield, gasification efficiency and carbon conversion reduced tar content (Huang et al., 2014b).

Wang et al. (2016) investigated CLG of biomass by thermodynamic method for higher H_2 and CO production and discussed the CLG reaction mechanism of biomass using Mn_2O_3 as OC. They found that at optimal thermodynamic conditions (1000°C and atmospheric pressure), H_2 and CO concentration was around 98.8% in syngas and coke formation on the Mn_2O_3 surface was not observed at high temperature (Wang et al., 2016).

CLG of pine sawdust tested with 10 kW capacity circulating FBR using Fe_2O_3/NiO bimetallic as an oxygen carrier by Wei et al. (2015). Fe-Ni bimetallic oxygen carrier exhibited constant activity with good resistivity against sintering. The syngas composition and gasification efficiency of biomass with Fe-Ni bimetallic oxygen carrier was found superior as compared to Fe_2O_3/Al_2O_3 OCs. The increase in yield of H_2 and CO and decreased trend in CO_2 and CH_4 were also observed with a rise in a temperature range of 760–910°C.

Ge et al. (2016b) employed natural hematite contains 83.2, 7.1, and 5.4 wt.% of Fe_2O_3, SiO_2 and Al_2O_3, respectively as OCs in a 25 kW_{th} interconnected fluidized bed reactor for rice husk CLG. The obtained results indicated that hematite mass percentage (40 wt.%) was found optimum to yield syngas (0.74 Nm^3/kg). However, carbon efficiency enhanced and the opposite trend in syngas yield was observed in the 40–60 wt.% range of natural hematite. They obtained a maximum of 0.74 Nm^3/kg syngas yield and carbon efficiency of about 89.2% with S/B ratio of 1 at 860°C. The study on biomass char CLG using iron ore modified NiO (10 wt.%) as OC in a fixed bed reactor demonstrated that the presence of OC increased CLG of char and obtained 55.7% carbon conversion. This can be due to a faster generation of CO as compared to other components because the lattice oxygen of OC partially oxidized carbon and rapidly consumed H_2 (Huang et al., 2014a).

Zeng et al. (2017) compared CLG performance of biomass self-moisture (BSM), dry biomass (DB) and biomass steam (BS) and demonstrated that gas yield increased to 1.1646 Nm^3/kg with moisture, whereas H_2/CO ratio increased to 0.744 mol/mol with steam and both favored reforming reaction of CH_4 and produced more H_2. It was observed that interaction between molecular oxygen of moisture or steam and lattice oxygen of OCs resulted in higher reactivity of biomass, but their influence was different in the processes. This was due to diverse water diffusion directions in gasification methods of biomass(Zeng et al., 2017). H_2 production and *in situ* removal of CO_2 from sawdust as biomass by CLG were performed two FBR using steam as a gasifying agent. They reported that at 580°C, and S/B ratio of 1.5, the gas composition of 71% H_2, 20% CH_4, 9% CO and no CO_2 was obtained. In this work, two FBR (i.e., bubbling bed and circulating bed) as gasifier and regenerator, respectively connected through cyclone were used for H_2 production. In a bubbling bed reactor, gasification rate and CO shift reactions can be augmented through *in situ* capturing CO_2 and were done by using CaO as sorbent, which absorbed CO_2 and converted into $CaCO_3$. This formed $CaCO_3$ is regenerated to CaO in a circulating bed reactor (Acharya et al., 2009).

Liu et al. (2018) performed CLG of microalgae using $CaFe_2O_4$ and $Ca_2Fe_2O_5$ as OCs in a fixed-bed reactor for syngas production. The reactivity of OCs and interactions between biomass and OC were discussed by the thermogravimetric analysis (TG) and thermogravimetric-Fourier transform infrared spectroscopy (TG-FTIR), respectively. They observed that steam addition had a positive impact on gasification efficiency and it enhanced from 83% to 92.5% with Fe_2O_3 and $Ca_2Fe_2O_5$ at 850°C.

Wu et al. (2018) observed that the addition of CaO not only abate the sintering phenomenon but also improved the reactivity of OC by forming $Ca_2Fe_2O_5$, which favored the production of syngas in CLG of biomass. In another study, four types of perovskites (e.g., $CaFeO_3$, $CaMnO_3$, $CaMn_{0.6}Fe_{0.4}O_3$ and $CaMn_{0.8}Fe_{0.2}O_3$) in a temperature range of 700–900°C and steam to char mass ratio between 3.6 and 8.4 was investigated in a fixed bed reactor for CLG of rice husk char to produce H_2-rich syngas (Liu et al., 2018). The results show that $CaMn_xFe_{1-x}O_3$ performed better than $CaFeO_3$ and $CaMnO_3$ and the highest 3.6 H_2/CO ratio, 1.5 mol/g gas yield and 1.4 mol/g CO_2 were achieved with $CaMn_{0.6}Fe_{0.4}O_3$ at 750°C and at steam to char mass ratio of 4.8. Herein, $CaMn_{0.6}Fe_{0.4}O_3$ played the vital role of OC and catalyst and required less energy as compared to other oxygen carrier material in gasification reaction. The CLG of biomass for high purity H_2 production with *in situ* CO_2 capture using demolition and concrete waste as sorbent source in an FBR at a temperature range of 650–1000°C and 2 MPa pressure examined by Moghtaderi et al. (2012). The results revealed a high H_2 containing syngas, 56.4% CO_2 removal efficiency combined with low cost, high attrition resistance and stability.

2.7 STEAM REFORMING PROCESS

Biomass is thermally degraded to yield bio-oil or syngas by gasification and pyrolysis processes. The product mixture contains very low H_2 content and is generally used as low-grade fuels. The bio-oil (60–75 wt.%) obtained from fast pyrolysis can be further catalytically promoted to produce H_2 as the main product using various reforming technologies, i.e., auto-thermal reforming, steam reforming, dry reforming, chemical looping reforming and partial oxidation (Balat et al., 2009; Wright et al., 2010). Table 2.2 depicts different reforming processes of H_2 production from biomass.

TABLE 2.2 Several Reforming Processes for H_2 Production

Process	Merits	Demerits	References
Autothermal reforming (ATR)	• N_2 free syngas • More favorable for CO_2 capture • Low operating temperature than partial oxidation	• High purity O_2 required • High production cost compared to steam reforming	Salkuyeh et al.(2017); Lavoie (2014); Xue et al. (2017)
Steam reforming (SR)	• Highest H_2/CO ratio obtained • Lower operating temperature than autothermal reforming • Industrially developed process	• High consumptions of energy and water, and CO_2 release	Chen et al. (2016); Salkuyeh et al. (2017); Lavoie (2014)
Partial oxidation reforming (PO)	• Minimum emission of SO_x • Little reuse of catalyst • Less slip of CH_4	• High purity O_2 required • High operating temperatures • Low H_2/CO ratio	Salkuyeh et al. (2017); Striūgas et al. (2012)
Chemical looping reforming (CLR)	• Low cost H_2 production • High thermal efficiency • Less CO_2 emission	• More industrial-scale studies are required	Salkuyeh et al. (2017); Luo et al. (2018)
Dry reforming (DR)	• Reduced greenhouse gas emissions	• High operating temperature • Low syngas ratio • More industrial-scale studies are required	Fu et al. (2016); Lavoie (2014)

Steam reforming (SR) is a widely accepted industrial process to produce high H_2 yield at the lowest operation cost. Hence, it is considered as valorization route for H_2 production using various parameters such as operating conditions reactors and catalysts. Apart from biomass feedstocks, other renewable raw materials such as methane (Meshksar et al., 2018), methanol

(González-Gil et al., 2016), ethanol (Isarapakdeetham et al., 2020), acetic acid (Yang et al., 2016) and glycerol (Zamzuri et al., 2017) have also been reported for H_2 production. Recently, Arregi et al. (2018) summarized various thermochemical routes to produce H_2 from aqueous and raw bio-oil obtained biomass and discussed influences of reactor configuration, steam/carbon (S/C) ratio, temperature, space-time, and catalysts for bio-oil steam reforming.

From thermodynamic studies, it was observed that high temperature and addition of steam favors H_2 production, while at high-pressure H_2 yield reduced using a low-cost nickel catalyst (Kim et al., 2015; Chitsazan et al., 2016; Charisiou et al., 2019). At elevated temperatures in SR, two effects were observed, such as suppression of WGS reaction led to reduced H_2 yield and increased decomposition of biomass (Rossetti et al., 2016). In general, a high S/C ratio led to enhance conversion and reduce carbon formation but at higher consumption of energy. In SR of hydrocarbons, C_xH_y* and OH* species are shaped from different species adsorbed on the catalyst surface. Furthermore, the formed C_xH_y*OH* and species react to generate CO and H_2, which subsequently convert to H_2 and CO_2 through WGS reaction. The following reactions are relevant in this regard (Rostrup-Nielsen et al., 2002):

$$CH_4 + 2^* \rightleftharpoons CH_3^* + H^* \tag{2.2}$$

$$CH_3^* + * \rightleftharpoons CH_2^* + H^* \tag{2.3}$$

$$CH_2^* + * \rightleftharpoons CH^* + H^* \tag{2.4}$$

$$CH^* + * \rightleftharpoons C^* + H^* \tag{2.5}$$

$$H_2O + 2^* \rightleftharpoons OH^* + H^* \tag{2.6}$$

$$OH^* + * \rightleftharpoons O^* + H^* \tag{2.7}$$

$$C^* + O^* \rightleftharpoons CO^* + * \tag{2.8}$$

$$CO^* \rightleftharpoons CO + * \tag{2.9}$$

$$2H^* \rightleftharpoons H_2 + 2^* \tag{2.10}$$

Carbon formation is an unavoidable event and depends on the operation conditions of the SR reaction. The carbon formed during SR leads to deactivating metal particles on the catalyst surface and may disintegrate support material. Carbon may be produced through the decomposition of CH_4 and C_nH_m and reduction of CO as follows (Hu and Lu, 2007; Roses et al., 2013). Various nickel and non-nickel catalyst and catalyst modification strategies for better improvement in steam reforming for H_2 production were discussed in detail by Dou et al. (2019).

Carbon decomposition:

$$CH_4 \rightleftharpoons C_s + H_2O \tag{2.11}$$

C_nH_m decomposition:

$$C_nH_m \rightleftharpoons nC_s + m/2H_2 \tag{2.12}$$

CO reduction:

$$CO + H_2 \rightleftharpoons C_s + H_2O \tag{2.13}$$

2.8 CHEMICAL LOOPING STEAM REFORMING

In chemical looping steam reforming (CLR), oxygen is transferred to fuel from the air with the help of OCs without direct access to fuel and air. A CLR system is made up of fuel and air reactors that are interconnected fluidized beds. An oxygen carrier continually circulates between these reactors. In the fuel reactor, CO_2 and H_2O are formed as the fuel oxidized by oxygen carriers and later on, WGS, and reforming reactions occur with steam. The oxygen carrier is regenerated within the air reactor. In general, a combination of reforming and partial oxidation processes of fuels is referred to as CLR for syngas. It is well known that partial oxidation and reforming are exothermic and endothermic processes, respectively. In this process to avoid 100% oxidation of fuel, the oxygen to fuel ratio was kept low.

Ortiz et al. (2011) investigated auto thermal operating conditions for CLRs with Ni-based oxygen carriers for maximum H_2 production via mass and heat balance. They reported that the molar ratio of oxygen to $CH_4 > 1.2$ was essential to attained auto thermal conditions and to produce maximum H_2. The major advantages of CLRs as compared to SR are a lower steam requirement, no additional energy requirement, no/little CO_2 emissions, less catalyst required, high reaction rates, and less carbon formation (Adanez et al., 2012).

Chemical looping water splitting (CLWS) is another promising technology to produce H_2 in two steps redox cycling. The two steps are reduction and oxidation, which occur by using metal oxides as the OCs. For the conversion of fuel, the OC provides lattice oxygen during the reduction step. The reduced oxygen carrier is re-oxidized in the oxidation step by water steam to form H_2. However, water-oxidized OCs might require oxidation by air in some cases (Zhao et al., 2016; Zhu et al., 2018). The reactions of CLWS can be illustrated as follows:

Reduction with fuel:

$$Me_xO_y + C_xH_yO_z \longrightarrow Me_xO_{y-1} + H_2 + CO \tag{2.14}$$

Oxidation:

$$H_2O + Me_xO_{y-1} \longrightarrow H_2 + Me_xO_y \tag{2.15}$$

Air oxidation:

$$Me_xO_{y-1} + O_2 \longrightarrow Me_xO_y \tag{2.16}$$

The selection and performance optimization of oxygen carriers are important for process efficiency. The development in synthesis, characterization, and performance in oxygen carrier materials reviewed by Protasova and Snijkers (2016). The oxygen carrier material used for H_2 production should have characteristics such as high oxygen-carrying capacity, the stability of multi-redox cycles, high reactivity with fuels, high resistance to agglomeration, cost-effectiveness, and environmentally friendliness (Protasova and Snijkers, 2016; Zhu et al., 2018).

2.9 SORPTION ENHANCED STEAM REFORMING

Sorption enhanced steam reforming (SESR) is another important technology to obtain low-cost H_2 with CO_2 capture (*in-situ*). This process is to shift the thermodynamic limit of the WGS reaction and enhances steam and feedstock conversion towards higher H_2. As compared to conventional steam reforming, SESR exhibits several advantages such as increased H_2 yield, decreases in the no of steps required for product separation and lowering reforming temperatures (Gil et al., 2015; Dou et al., 2016; Xie et al., 2017). The *in situ* CO_2 elimination in the SESR reaction can be express as follows:

Reforming:

$$C_xH_yO_z \longrightarrow H_2 + CO \tag{2.17}$$

Water-gas shift:

$$CO + H_2O \longrightarrow H_2 + CO_2 \tag{2.18}$$

CO_2 sorption:

$$MeO + CO_2 \longrightarrow MeCO_3 \tag{2.19}$$

In CO_2 sorption, MeO can be a solid sorbent such as CaO, Li_2ZrO_3, etc. SESR efficiency strongly depends on sorbent performance for *in situ* CO_2 capture by and fast rates of sorption and regeneration for WGS, steam reforming, CO_2 regeneration, and capture reactions. Various sorbents including CaO, MgO, hydrotalcite, Li_2ZrO_3, multi-functional sorbent catalyst minerals have been reviewed for CO_2 capture (Dou et al., 2016). Dou et al. (2009) performed H_2 production by SESR of glycerol between 400°C and 700°C at atmospheric pressure with a fixed bed reactor using calcined dolomite sorbent and results showed that H_2 productivity was augmented with rising temperature. The maximum 97% H_2 purity was attained with the longest CO_2 breakthrough time at 500°C.

Fermoso et al. (2012) employed Ni/Co as catalyst and dolomite (calcined) as CO_2 sorbent in a single-stage SESR reaction to crude glycerol. They concluded that 88% yield and 99.7% purity of H_2 were obtained at 550°C, 1 atm pressure, S/C of 3 and spatial time of 1.09 h. The *in situ* CO_2 removal by calcined dolomite sorbent increased the rates for WGS and reforming reactions. Moreover, they also observed that catalytic activity remained intact in three repeated cycles.

Dewoolkar and Vaidya (2016) used Ni-incorporated four different cationic modified (e.g., Cu^{2+}, Mg^{2+}, Zn^{2+}, and Ca^{2+}) based hydrotalcite hybrid materials for SESR of ethanol and studied the effects operating parameters like S/C ratio, temperature, and mass fraction of sorbent on H_2 production. The used materials (hybrid) exhibited CO_2 adsorption capacities as follows: Cu > Mg > Zn > Ca and reported that Cu-based hybrid material produced H_2 (99 mol%) with 1.2 mol CO_2/kg sorbent adsorption capacity at reaction conditions (temperature: 300°C, pressure: 1 atm, S/C ratio: 5 and breakthrough time: 70 min). Cu-based hybrid material remained stable up to 18 cycles and that of Mg-based, which was stable up to 21 cycles (Dewoolkar and Vaidya, 2016). Cunha et al. (2013) studied CuMgAl hybrid materials for SESR of ethanol and reported obtaining over 90% H_2 purity. The activities of CuMgAl materials remained intact throughout 4 h on stream in temperature between 200°C and 600°C.

A bi-functional catalyst owing to good contact between catalyst and sorbent can reduce the complexity of the process by integrating properties of sorbent and catalyst, which in turn improves performance has used in the SESR process. Dang et al. (2016) synthesized a bi-functional catalyst Co-CaO-$Ca_{12}Al_{14}O_{33}$ by co-precipitation method for SESR of glycerol, which increased H_2 concentration to 96.4% at 525°C in pre breakthrough stage. The H_2 concentration remains constant up to 50 reaction-calcination cycles. Hu et al. (2017) examined the SESR of acetic acid using Ni/$Ce_xZr_{1-x}O_2$-CaO catalyst synthesized by the sol-gel method at 550°C and an S/C ratio of 4. They obtained 98% purity of H_2 during the pre-breakthrough time, which was sustained up to 15 cycles.

Yancheshmeh et al. (2017) performed SESR of glycerol to produce H_2 with high purity using $Ca_9A_{16}O_{18}$-aO/xNiO (x = 15, 20, and 25 wt.%) and $Ca_9A_{16}O_1$-CaO/20NiO-yCeO_2 (y = 5, 10, and 15 wt.%) bi-functional catalyst. They observed that the sintering of CaO and Ni and coke formation not only decreased the reactivity of all $Ca_9A_{16}O_{18}$-CaO/xNiO catalysts but also reduced CO_2 removal efficiency. The highest H_2 purity (98%) was obtained with $Ca_9A_{16}O_{18}$-CaO/20NiO. The addition of CeO_2 to $Ca_9A_{16}O_{18}$-CaO/20NiO significantly improved material stability during cyclic operation. This was due to higher interaction of NiO and CeO_2 and CeO_2, which controlled active sites of Ni, which led to enhanced coke gasification and declined coke formation. A maximum 91% H_2 yield with 98% purity was obtained with $Ca_9A_{16}O_{18}$-CaO/20NiO-10 CeO_2 catalyst over 20 regeneration cycles.

Wang et al. (2015) investigated the SESR of glycerol to produce H_2 over a multifunctional catalyst (Ni-based) with CO_2 capture (*in situ*) using a fixed-bed reactor with CaO sorbent. The CO_2 sorption capability of a multifunctional catalyst was determined by isothermal and non-isothermal experiments. Jiang et al. (2017) used a series of multifunctional catalysts (i.e., Ni-CaO-MMT, and MMT-Montmorillonite) with a different ration of Ca/MMT synthesized by an ultrasound-assisted cation-exchange method for SESR of glycerol and obtained characterization results confirmed the strong interactions between metal and support with a uniform dispersion of Ca and Ni, is interconnected to internment effect of MMT structure.

Recently, Ghungrud and Vaidya (2020) used 10 wt.% Co and ZrO_2, MgO, and CeO_2 modified based Ca sorbent as a multifunctional catalyst for

SESR of ethanol. It was observed that Ce and Zr hybrid materials exhibited superior CO_2 adsorption capacity together with improved pre-breakthrough time than Mg-based material. At 500°C, Co-CaO/CeO_2 displayed 5.6 mol CO_2/kg adsorption ability and 45 min breakthrough time and produced 80 mol% H_2, while Co-CaO/ZrO_2 displayed 3.8 mol CO_2/kg sorbent adsorption capacity with a breakthrough time of 30 min with about 75% H_2. Co-CaO/MgO demonstrated the lowest adsorption capacity of 1.5 mol CO_2/kg sorbent combined with a 15 min breakthrough time and producing H_2 (70 mol%) (Ghungrud and Vaidya, 2020).

Arstad et al. (2012) used Ni/$NiAl_2O_4$ and calcined dolomite as catalyst and sorbent, respectively in methane steam reforming enhanced by sorbent for continuous H_2 production with circulating fluidized bed (CFB) reactor at 575°C, atmospheric pressure and steam/methane ratio of 4. The results demonstrated that CH_4 conversion in the CFB reactor was drastically lower than the fixed bed reactor. This indicated that the CFB system required a high catalyst amount as compared to a fixed bed system.

Dou et al. (2013) used employed two moving bed reactors with a continuous flow of sorbent and catalyst in SESR of glycerol for H_2 generation and were able to produce high purity H_2 by integrating various process units of WGS, SR, and elimination of CO_2 (*in situ*) in a single stage. The requirement WGS catalyst was eliminated under certain operating conditions. Lysikov et al. (2015) investigated SESR of bio-ethanol in a dual fixed bed reactor using Ni as a catalyst and CaO as sorbent. The obtained results showed that 50 L/min continuous H_2 production and 97–99 vol% H_2 purity were obtained and maintained in both, temperature, and pressure swing adsorption, and the system worked efficiently for greater than 250 cycles.

2.10 SORPTION ENHANCED CHEMICAL LOOPING STEAM REFORMING

Sorption enhanced chemical looping steam reforming (SE-CLSR) is an innovative approach to produce H_2 by combing CLSR with CO_2 removal (*in-situ*) through solid sorbents. This process can reduce the overall heat demands by combined exothermic and endothermic reactions in both stages. The oxygen carrier re-oxidation and sorbent regeneration in air reactor and CaO carbonation and reforming in the fuel reactor and can be considered as a low-cost energy-efficient process (Antzara et al., 2016; Abbas et al., 2017).

Hafizi et al. (2016) studied the SE-CLSR process using $Fe_2O_3/MgAl_2O_4$ and Fe_2O_3/Al_2O_3 as the OCs with 22 wt.% of Fe_2O_3 in a fixed bed reactor. They investigated the activity of industrial, prepared, and Ce-promoted CaO for CO_2 removal and found that the addition of Ce enhanced the structural properties of CaO and improved CO_2 sorption capacity owing to the high surface area as compared to prepared and industrial CaO sorbent. The obtained results revealed that the addition of Mg in OC structure inhibits $FeAl_2O_4$ formation by generating $MgAl_2O_4$. The average methane conversion in the SE-CLSR process was found lower than the CLR process at all temperatures. This was due to the mass transfer resistance between the oxygen carrier and sorbent material.

Spragg et al. (2018) performed steam reforming, CLSR, and SE-CLSR process using Ni/NiO as OC and CaO as sorbent, and compared their performance for the generation of H_2 from the bio-oil surrogate mixture. They found that the highest H_2 purity (99.7 mol%) and a low net energy balance were achieved at 500°C with SE-CLSR. The results revealed that carbon can be diminished by an increased amount of sorbent or oxygen carrier material at an S/C of 1.

Dou et al. (2018) conducted SE-CLSR of ethanol to investigate the additional effect of CaO-based sorbent for capture of CO_2 in a fixed-bed reactor (alternating) with NiO/Al_2O_3 as OC at 600°C and atmospheric pressure with S/C of 3. They reported that the highest selectivity of H_2 and conversion were obtained with a Ca/Ni ratio of 3. A further enhancement in the Ca/Ni molar ratio led to reduced H_2 production due to coke deposition and dilution of oxygen carrier by high sorbent amount.

Adiya et al. (2017) conducted SE-CLSR of Shell gas (Marcellus) (roughly up to 80% CH_4, >20% hydrocarbons, CO_2 and N_2) with NiO as both catalyst and oxygen transfer material (OTM) and CaO as a sorbent to produce H_2 based on thermodynamic equilibrium analysis. They obtained 49% H_2 yield and 98% H_2 selectivity in the SE-CLSR process, which was higher than conventional steam reforming (24.3% H_2 yield and 65.4% H_2 selectivity) at S/C of 3 and 527°C. The minimum energy of 159 kJ/mol was needed in steam reforming to produce H_2 at 527°C with S/C of 3, which drastically reduced to 34 kJ (without sorbent regeneration) and 92 kJ (with sorbent regeneration at 897°C) per mol of H_2 in SE-CLSR at same operation condition.

Ni et al. (2017) employed Ni supported on $NiAl_2O_4$ and ZrO_2 as OTMs and Li_2ZrO_3 based sorbent to produced high purity H_2 from SE-CLSR of glycerol continuous flow packed bed reactor at 550°C with S/C of 3 and reported that H_2 concentration (> 90%) was achieved with both OTMs, which remains stable during 10 cycles indicated coke deposition effect on the OTMs surface could successfully reduce in SE-CLSR process.

2.11 CONCLUSIONS

Continuous exploitation of fossil fuels has led to enhance the magnitude of environmental and economic complications, particularly greenhouse gas emissions. This boosted interest and research on the production of valuable chemicals from renewable resources. Biomass-derived products (chemicals and fuels) not only play a crucial role in declining CO_2 emissions but also considered as a source for sustainability and energy competitiveness. Gasification technologies can produce H_2 via biomass conversion. However, the diversities in biomass materials, i.e., particle size, moisture content, ash content, biomass types require special attention.

Although, biomass conversion offers several advantages for H_2 production due to various technical challenges related to conversion methods is not commercialized. For instance, SCWG can be operated with biomass moisture content (>30%), but higher energy demand and problems in the feeding of biomass hindered its applicability at a large scale. Other thermochemical processes such as gasification, steam reforming, and pyrolysis have a high capability of biomass usage for industrial execution. However, owing to the thermodynamic limitations of reversible WGS reactions, these processes are found to be inadequate in the conversion of biomass feedstock and H_2 efficiency. The CLSR and SESR technologies are proven important techniques for *in-situ* CO_2 removal from product mixture and production of high purity H_2. These techniques do not only produce high purity H_2 but also consume less energy. However, issues such as sintering, CO_2 sorption capacities, stability, and high adsorption/desorption rates at reduced regeneration temperatures of sorbent are needed to be addressed.

It has been observed that H_2 production enhances in the presence of the catalyst. However, high selectivity and purity of H_2 are some major concerns for the technologies. The strong metal-support interaction, surface oxygen mobility together with *in-situ* CO_2 removal are major challenges associated with catalyst design. Apart from these, stability, attrition, resistivity, poisoning of sorbents and catalysts are also important for their performance and lifecycles. A better understanding of these parameters could help to design novel sorbents and catalysts to achieve higher reactivity of the existing reactions. Hence, the research focus should be on the production of H_2-rich gas to minimize CO_2 emission through the optimization of process parameters of thermo-catalytic conversion of biomass.

KEYWORDS

- **biomass**
- **catalysts**
- **dual fluidized bed reactor**
- **equivalence ratio**
- **fluidized bed reactor**
- **gasification**
- **hydrogen**
- **reforming**
- **steam-to-biomass ratio**

REFERENCES

Abbas, S. Z., Dupont, V., & Mahmud, T., (2017). Modeling of high purity H_2 production via sorption enhanced chemical looping steam reforming of methane in a packed bed reactor. *Fuel, 202*, 271–286.

Abuadala, A., & Dincer, I., (2012). A review on biomass-based hydrogen production and potential applications: A review on biomass-based hydrogen production and applications. *Int. J. Energy Res., 36*, 415–455.

Acharya, B., Dutta, A., & Basu, P., (2009). Chemical-looping gasification of biomass for hydrogen-enriched gas production with in-process carbon dioxide capture. *Energy Fuels, 23*, 5077–5083.

Adanez, J., Abad, A., Garcia-Labiano, F., Gayan, P., & De Diego, L. F., (2012). Progress in chemical-looping combustion and reforming technologies. *Prog. Energy Combust. Sci., 38*, 215–282.

Adiya, Z. I. S. G., Dupont, V., & Mahmud, T., (2017). Chemical equilibrium analysis of hydrogen production from shale gas using sorption enhanced chemical looping steam reforming. *Fuel Process. Technol., 159*, 128–144.

Ahmad, A. A., Zawawi, N. A., Kasim, F. H., Inayat, A., & Khasri, A., (2016). Assessing the gasification performance of biomass: A review on biomass gasification process conditions, optimization and economic evaluation. *Renew. Sust. Energy Rev., 53*, 1333–1347.

Ahmadi, P., Dincer, I., & Rosen, M. A., (2013). Development and assessment of an integrated biomass-based multi-generation energy system. *Energy, 56*, 155–166.

Ahmed, I. I., & Gupta, A. K., (2011). Kinetics of woodchips char gasification with steam and carbon dioxide. *Appl. Energy, 88*, 1613–1619.

Anca-Couce, A., (2016). Reaction mechanisms and multi-scale modelling of lignocellulosic biomass pyrolysis. *Prog. Energy Combust. Sci., 53*, 41–79.

Antzara, A., Heracleous, E., & Lemonidou, A. A., (2016). Energy efficient sorption enhanced-chemical looping methane reforming process for high-purity H_2 production: Experimental proof-of-concept. *Appl. Energy, 180*, 457–471.

Arena, U., (2012). Process and technological aspects of municipal solid waste gasification: A review. *Waste Manag., 32*, 625–639.

Arregi, A., Amutio, M., Lopez, G., Bilbao, J., & Olazar, M., (2018). Evaluation of thermochemical routes for hydrogen production from biomass: A review. *Energy Convers. Manag., 165*, 696–719.

Arstad, B., Prostak, J., & Blom, R., (2012). Continuous hydrogen production by sorption enhanced steam methane reforming (SE-SMR) in a circulating fluidized bed reactor: Sorbent to catalyst ratio dependencies. *Chem. Eng. J., 189, 190*, 413–421.

Artetxe, M., Lopez, G., Amutio, M., Elordi, G., Olazar, M., & Bilbao, J., (2010). Operating conditions for the pyrolysis of poly (ethylene terephthalate) in a conical spouted-bed reactor. *Ind. Eng. Chem. Res., 49*, 2064–2069.

Ates, F., Putun, A., & Putun, E., (2006). Pyrolysis of two different biomass samples in a fixed-bed reactor combined with two different catalysts. *Fuel, 85*, 1851–1859.

Azadi, P., & Farnood, R., (2011). Review of heterogeneous catalysts for sub- and supercritical water gasification of biomass and wastes. *Int. J. Hydrogen Energy, 36*, 9529–9541.

Balat, M., (2008). Hydrogen-rich gas production from biomass via pyrolysis and gasification processes and effects of catalyst on hydrogen yield. *Energy Sourc. Part A, 30*, 552–564.

Balat, M., Balat, M., Kırtay, E., & Balat, H., (2009). Main routes for the thermo-conversion of biomass into fuels and chemicals Part 1: Pyrolysis systems. *Energy Convers. Manag., 50*, 3147–3157.

Barisano, D., Canneto, G., Nanna, F., Alvino, E., Pinto, G., Villone, A., Carnevale, Met al., (2016). Steam/oxygen biomass gasification at pilot scale in an internally circulating bubbling fluidized bed reactor. *Fuel Process. Technol., 141*, 74–81.

Basu, P., (2010a). Biomass characteristics. In: *Biomass Gasification Design Handbook* (pp. 27–63). Elsevier.

Basu, P., (2010b). Design of biomass gasifiers. In: *Biomass Gasification Design Handbook* (pp. 167–228). Elsevier.

Basu, P., (2010c). Gasification theory and modeling of gasifiers. In: *Biomass Gasification Design Handbook* (pp. 117–165). Elsevier.

Bellouard, Q., Abanades, S., & Rodat, S., (2017a). Biomass gasification in an innovative spouted-bed solar reactor: Experimental proof of concept and parametric study. *Energy Fuels, 31*, 10933–10945.

Bellouard, Q., Abanades, S., Rodat, S., & Dupassieux, N., (2017b). Solar thermochemical gasification of wood biomass for syngas production in a high-temperature continuously fed tubular reactor. *Int. J. Hydrogen Energy, 42*, 13486–13497.

Burhenne, L., Messmer, J., Aicher, T., & Laborie, M. P., (2013). The effect of the biomass components lignin, cellulose and hemicellulose on TGA and fixed bed pyrolysis. *J. Anal. Appl. Pyrolysis, 101*, 177–184.

Charisiou, N. D., Papageridis, K. N., Tzounis, L., Sebastian, V., Hinder, S. J., Baker, M. A., AlKetbi, M., et al., (2019). Ni supported on CaO-MgO-Al_2O_3 as a highly selective and stable catalyst for H_2 production via the glycerol steam reforming reaction. *Int. J. Hydrogen Energy, 44*, 256–273.

Chen, G., Yao, J., Liu, J., Yan, B., & Shan, R., (2016). Biomass to hydrogen-rich syngas via catalytic steam reforming of bio-oil. *Renew. Energy, 91*, 315–322.

Chen, Y., Guo, L., Cao, W., Jin, H., Guo, S., & Zhang, X., (2013). Hydrogen production by sewage sludge gasification in supercritical water with a fluidized bed reactor. *Int. J. Hydrogen Energy, 38*, 12991–12999.

Cheng, Y. T., Jae, J., Shi, J., Fan, W., & Huber, G. W., (2012). Production of renewable aromatic compounds by catalytic fast pyrolysis of lignocellulosic biomass with bifunctional Ga/ZSM-5 catalysts. *Angew. Chem. Int. Ed., 51*, 1387–1390.

Chitsazan, S., Sepehri, S., Garbarino, G., Carnasciali, M. M., & Busca, G., (2016). Steam reforming of biomass-derived organics: Interactions of different mixture components on Ni/Al_2O_3 based catalysts. *Appl. Catal. B: Environ., 187*, 386–398.

Corella, J., Toledo, J. M., & Molina, G., (2007). A review on dual fluidized-bed biomass gasifiers. *Ind. Eng. Chem. Res., 46*, 6831–6839.

Correa, C. R., & Kruse, A., (2018). Supercritical water gasification of biomass for hydrogen production: Review. *J. Supercrit. Fluids, 133*, 573–590.

Cunha, A. F., Wu, Y. J., Santos, J. C., & Rodrigues, A. E., (2013). Sorption enhanced steam reforming of ethanol on hydrotalcite-like compounds impregnated with active copper. *Chem. Eng. Res. Des., 91*, 581–592.

D'Jesús, P., Boukis, N., Kraushaar-Czarnetzki, B., & Dinjus, E., (2006). Gasification of corn and clover grass in supercritical water. *Fuel, 85*, 1032–1038.

Dang, C., Yu, H., Wang, H., Peng, F., & Yang, Y., (2106). A bi-functional Co-CaO-$Ca_{12}Al_{14}O_{33}$ catalyst for sorption-enhanced steam reforming of glycerol to high-purity hydrogen. *Chem. Eng. J., 286*, 329–338.

Dayton, D., (2002). *A Review of the Literature on Catalytic Biomass Tar Destruction*. National Renewable Energy Laboratory, U.S. Department of Energy Laboratory report. https://doi.org/10.2172/15002876.

De Andrés, J. M., Narros, A., & Rodríguez, M. E., (2011). Behavior of dolomite, olivine and alumina as primary catalysts in air-steam gasification of sewage sludge. *Fuel, 90*, 521–527.

De Lasa, H., Salaices, E., Mazumder, J., & Lucky, R., (2011). Catalytic steam gasification of biomass: Catalysts, thermodynamics and kinetics. *Chem. Rev., 111*, 5404–5433.

Delgado, J., Aznar, M. P., & Corella, J., (1996). Calcined dolomite, magnesite, and calcite for cleaning hot gas from a fluidized bed biomass gasifier with steam: Life and usefulness. *Ind. Eng. Chem. Res., 35*, 3637–3643.

Demirbaş, A., (2002). Hydrogen production from biomass by the gasification process. *Energy Sourc., 24*, 59–68.

Devi, L., Ptasinski, K. J., Janssen, F. J. J. G., Van, P. S. V. B., Bergman, P. C. A., & Kiel, J. H. A., (2005). Catalytic decomposition of biomass tars: Use of dolomite and untreated olivine. *Renew. Energy, 30*, 565–587.

Dewoolkar, K. D., & Vaidya, P. D., (2016). Tailored hydrotalcite-based hybrid materials for hydrogen production via sorption-enhanced steam reforming of ethanol. *Int. J. Hydrogen Energy, 41*, 6094–6106.

Diblasi, C., (2008). Modeling chemical and physical processes of wood and biomass pyrolysis. *Prog. Energy Combust. Sci., 34*, 47–90.

Din, Z. U., & Zainal, Z. A., (2016). Biomass integrated gasification-SOFC systems: Technology overview. *Renewable and Sustainable Energy Reviews, 53*, 1356–1376.

Dou, B., Dupont, V., Rickett, G., Blakeman, N., Williams, P. T., Chen, H., Ding, Y., & Ghadiri, M., (2009). Hydrogen production by sorption-enhanced steam reforming of glycerol. *Bioresour. Technol., 100*, 3540–3547.

Dou, B., Wang, C., Chen, H., Song, Y., & Xie, B., (2013). Continuous sorption-enhanced steam reforming of glycerol to high-purity hydrogen production. *Int. J. Hydrogen Energy, 38*, 11902–11909.

Dou, B., Wang, C., Song, Y., Chen, H., Jiang, B., Yang, M., & Xu, Y., (2016). Solid sorbents for in-situ CO_2 removal during sorption-enhanced steam reforming process: A review. *Renew. Sust. Energy Rev., 53*, 536–546.

Dou, B., Zhang, H., Cui, G., Wang, Z., Jiang, B., Wang, K., Chen, H., & Xu, Y., (2018). Hydrogen production by sorption-enhanced chemical looping steam reforming of ethanol in an alternating fixed-bed reactor: Sorbent to catalyst ratio dependencies. *Energy Convers. Manag., 155*, 243–252.

Dou, B., Zhang, H., Song, Y., Zhao, L., Jiang, B., He, M., Ruan, C., Chen, H., & Xu, Y., (2019). Hydrogen production from the thermochemical conversion of biomass: Issues and challenges. *Sustain. Energy Fuels, 3*, 314–342.

Duman, G., & Yanik, J., (2017). Two-step steam pyrolysis of biomass for hydrogen production. *Int. J. Hydrogen Energy, 42*, 17000–17008.

Elif, D., & Nezihe, A., (2016). Hydrogen production by supercritical water gasification of fruit pulp in the presence of Ru/C. *Int. J. Hydrogen Energy, 41*, 8073–8083.

Erkiaga, A., Lopez, G., Amutio, M., Bilbao, J., & Olazar, M., (2013). Steam gasification of biomass in a conical spouted bed reactor with olivine and γ-alumina as primary catalysts. *Fuel Process. Technol., 116*, 292–299.

Erkiaga, A., Lopez, G., Amutio, M., Bilbao, J., & Olazar, M., (2014). Influence of operating conditions on the steam gasification of biomass in a conical spouted bed reactor. *Chem. Eng. J., 237*, 259–267.

Farzad, S., Mandegari, M. A., & Görgens, J. F., (2016). A critical review on biomass gasification, co-gasification, and their environmental assessments. *Biofuel Res. J., 3*, 483–495.

Feng, W., Van, D. K. H. J., & De Swaan, A. J., (2004). Biomass conversions in subcritical and supercritical water: Driving force, phase equilibria, and thermodynamic analysis. *Chem. Eng. Process.: Process Intens., 43*, 1459–1467.

Feng, Y., Xiao, B., Goerner, K., & Naidu, R., (2011). Influence of particle size and temperature on gasification performance. *AMR, 281*, 78–83.

Fermoso, J., He, L., & Chen, D., (2012). Production of high purity hydrogen by sorption enhanced steam reforming of crude glycerol. *Int. J. Hydrogen Energy, 37*, 14047–14054.

Fermoso, J., Stevanov, C., Moghtaderi, B., Arias, B., Pevida, C., Plaza, M. G., Rubiera, F., & Pis, J. J., (2009). High-pressure gasification reactivity of biomass chars produced at different temperatures. *J. Anal. Appl. Pyrolysis, 85*, 287–293.

Fremaux, S., Beheshti, S. M., Ghassemi, H., & Shahsavan-Markadeh, R., (2015). An experimental study on hydrogen-rich gas production via steam gasification of biomass in a research-scale fluidized bed. *Energy Convers. Manag., 91*, 427–432.

Fu, M., Qi, W., Xu, Q., Zhang, S., & Yan, Y., (2016). Hydrogen production from bio-oil model compounds dry (CO_2) reforming over Ni/Al_2O_3 catalyst. *Int. J. Hydrogen Energy, 41*, 1494–1501.

Ge, H., Guo, W., Shen, L., Song, T., & Xiao, J., (2016a). Experimental investigation on biomass gasification using chemical looping in a batch reactor and a continuous dual reactor. *Chem. Eng. J., 286*, 689–700.

Ge, H., Guo, W., Shen, L., Song, T., & Xiao, J., (2016b). Biomass gasification using chemical looping in a 25 kWth reactor with natural hematite as oxygen carrier. *Chem. Eng. J., 286*, 174–183.

Ghungrud, S. A., & Vaidya, P. D., (2020). Improved hydrogen production from sorption-enhanced steam reforming of ethanol (SESRE) using multifunctional materials of cobalt catalyst and Mg-, Ce-, and Zr-modified CaO Sorbents. *Ind. Eng. Chem. Res., 59*, 693–703.

Gil, J., & Corella, J., (1999). Biomass gasification in atmospheric and bubbling fluidized bed: Effect of the type of gasifying agent on the product distribution. *Biomass Bioenergy, 17*, 389–403.

Gil, M. V., Fermoso, J., Rubiera, F., & Chen, D., (2015). H_2 production by sorption enhanced steam reforming of biomass-derived bio-oil in a fluidized bed reactor: An assessment of the effect of operation variables using response surface methodology. *Catal. Today, 242*, 19–34.

González-Gil, R., Herrera, C., Larrubia, M. Á., Kowalik, P., Pieta, I. S., & Alemany, L. J., (2016). Hydrogen production by steam reforming of DME over Ni-based catalysts modified with vanadium. *Int. J. Hydrogen Energy, 41*, 19781–19788.

Göransson, K., Söderlind, U., & Zhang, W., (2001). Experimental test on a novel dual fluidized bed biomass gasifier for synthetic fuel production. *Fuel, 90*, 1340–1349.

Göransson, K., Söderlind, U., He, J., & Zhang, W., (2011). Review of syngas production via biomass DFBGs. *Renew. Sust. Energy Rev., 15*, 482–492.

Guo, Y., Wang, S. Z., Xu, D. H., Gong, Y. M., Ma, H. H., & Tang, X. Y., (2010). Review of catalytic supercritical water gasification for hydrogen production from biomass. *Renew. Sust. Energy Rev., 14*, 334–343.

Hafizi, A., Rahimpour, M. R., & Hassanajili, S., (2016). High purity hydrogen production via sorption enhanced chemical looping reforming: Application of $22Fe_2O_3/MgAl_2O_4$ and 22 Fe_2O_3/Al_2O_3 as oxygen carriers and cerium promoted CaO as CO_2 sorbent. *Appl. Energy, 169*, 629–641.

Hamad, M. A., Radwan, A. M., Heggo, D. A., & Moustafa, T., (2016). Hydrogen rich gas production from catalytic gasification of biomass. *Renew. Energy, 85*, 1290–1300.

Hanping, C., Bin, L., Haiping, Y., Guolai, Y., & Shihong, Z., (2008). Experimental investigation of biomass gasification in a fluidized bed reactor. *Energy Fuels, 22*, 3493–3498.

Heidenreich, S., & Foscolo, P. U., (2015). New concepts in biomass gasification. *Prog. Energy Combust. Sci., 46*, 72–95.

Hernández, J. J., Aranda, G., Barba, J., & Mendoza, J. M., (2012). Effect of steam content in the air-steam flow on biomass entrained flow gasification. *Fuel Process. Technol., 99*, 43–55.

Hernández, J. J., Aranda-Almansa, G., & Bula, A., (2010). Gasification of biomass wastes in an entrained flow gasifier: Effect of the particle size and the residence time. *Fuel Process. Technol., 91*, 681–692.

Hossain, M. Z., & Charpentier, P. A., (2015). Hydrogen production by gasification of biomass and opportunity fuels. In: *Compendium of Hydrogen Energy* (pp. 137–175). Elsevier.

Hossain, M. Z., Chowdhury, M. B. I., Jhawar, A. K., & Charpentier, P. A., (2017). Supercritical water gasification of glucose using bimetallic aerogel $Ru\text{-}Ni\text{-}Al_2O_3$ catalyst for H_2 production. *Biomass Bioenergy, 107*, 39–51.

Hosseini, S. E., Abdul, W. M., Jamil, M. M., Azli, A. A. M., & Misbah, M. F., (2015). A review on biomass-based hydrogen production for renewable energy supply: Biomass-based hydrogen production for renewable energy supply. *Int. J. Energy Res., 39*, 1597–1615.

Hu, G., Xu, S., Li, S., Xiao, C., & Liu, S., (2006). Steam gasification of apricot stones with olivine and dolomite as downstream catalysts. *Fuel Process. Technol., 87*, 375–382.

Hu, R., Li, D., Xue, H., Zhang, N., Liu, Z., & Liu, Z., (2017). Hydrogen production by sorption-enhanced steam reforming of acetic acid over Ni/$Ce_xZr_{1-x}O_2$-CaO catalysts. *Int. J. Hydrogen Energy, 42*, 7786–7797.

Hu, X., & Lu, G., (2007). Investigation of steam reforming of acetic acid to hydrogen over Ni-Co metal catalyst. *J. Mol. Catal. A: Chem., 261*, 43–48.

Huang, B. S., Chen, H. Y., Chuang, K. H., Yang, R. X., & Wey, M. Y., (2012). Hydrogen production by biomass gasification in a fluidized-bed reactor promoted by an Fe/CaO catalyst. *Int. J. Hydrogen Energy, 37*, 6511–6518.

Huang, Z., He, F., Feng, Y., Zhao, K., Zheng, A., Chang, S., Wei, G., Zhao, Z., & Li, H., (2014a). Biomass char direct chemical looping gasification using NiO-modified iron ore as an oxygen carrier. *Energy Fuels, 28*, 183–191.

Huang, Z., He, F., Zhao, K., Feng, Y., Zheng, A., Chang, S., Zhao, Z., & Li, H., (2014b). Natural iron ore as an oxygen carrier for biomass chemical looping gasification in a fluidized bed reactor. *J. Therm. Anal. Calorim., 116*, 1315–1324.

Huang, Z., He, F., Zheng, A., Zhao, K., Chang, S., Li, X., Li, H., & Zhao, Z., (2013). Thermodynamic analysis and synthesis gas generation by chemical-looping gasification of biomass with nature hematite as oxygen carriers. *J. Sustain. Bioenergy Syst., 3*, 33–39.

Huang, Z., Zhang, Y., Fu, J., Yu, L., Chen, M., Liu, S., He, F., Chen, D., Wei, G., Zhao, K., Zheng, A., Zhao, Z., & Li, H., (2016). Chemical looping gasification of biomass char using iron ore as an oxygen carrier. *Int. J. Hydrogen Energy, 41*, 17871–17883.

Huda, A. S. N., Mekhilef, S., & Ahsan, A., (2014). Biomass energy in Bangladesh: Current status and prospects. *Renew. Sust. Energy Rev., 30*, 504–517.

Imran, A., Bramer, E., Seshan, K., & Brem, G., (2016). Catalytic flash pyrolysis of biomass using different types of zeolite and online vapor fractionation. *Energies, 9*, 187.

Iovane, P., Donatelli, A., & Molino, A., (2013). Influence of feeding ratio on steam gasification of palm shells in a rotary kiln pilot plant. Experimental and numerical investigations. *Biomass Bioenergy, 56*, 423–431.

Isarapakdeetham, S., Kim-Lohsoontorn, P., Wongsakulphasatch, S., Kiatkittipong, W., Laosiripojana, N., Gong, J., & Assabumrungrat, S., (2020). Hydrogen production via chemical looping steam reforming of ethanol by Ni-based oxygen carriers supported on CeO_2 and La_2O_3 promoted Al_2O_3. *Int. J. Hydrogen Energy, 45*, 1477–1491.

Jiang, B., Dou, B., Wang, K., Zhang, C., Li, M., Chen, H., & Xu, Y., (2017). Sorption enhanced steam reforming of biodiesel by-product glycerol on Ni-CaO-MMT multifunctional catalysts. *Chem. Eng. J., 313*, 207–216.

Kan, T., Strezov, V., & Evans, T. J., (2016). Lignocellulosic biomass pyrolysis: A review of product properties and effects of pyrolysis parameters. *Renew. Sust. Energy Rev., 57*, 1126–1140.

Kaushal, P., & Tyagi, R., (2012). Steam assisted biomass gasification-an overview. *Can. J. Chem. Eng., 90*, 1043–1058.

Kihedu, J. H., Yoshiie, R., & Naruse, I., (2016). Performance indicators for air and air-steam auto-thermal updraft gasification of biomass in packed bed reactor. *Fuel Process. Technol., 141*, 93–98.

Kim, D., Kwak, B. S., Min, B. K., & Kang, M., (2015). Characterization of Ni and W co-loaded SBA-15 catalyst and its hydrogen production catalytic ability on ethanol steam reforming reaction. *Appl. Surf. Sci., 332*, 736–746.

Knight, R. A., (2000). Experience with raw gas analysis from pressurized gasification of biomass. *Biomass Bioenergy, 18*, 67–77.

Koppatz, S., Pfeifer, C., & Hofbauer, H., (2011). Comparison of the performance behavior of silica sand and olivine in a dual fluidized bed reactor system for steam gasification of biomass at pilot plant scale. *Chem. Eng. J., 175*, 468–483.

Lavoie, J. M., (2014). Review on dry reforming of methane, a potentially more environmentally friendly approach to the increasing natural gas exploitation. *Front. Chem., 2*, 81.

Li, J., Yin, Y., Zhang, X., Liu, J., & Yan, R., (2009). Hydrogen-rich gas production by steam gasification of palm oil wastes over supported tri-metallic catalyst. *Int. J. Hydrogen Energy, 34*, 9108–9115.

Liu, C., Wang, W., & Chen, D., (2018). Hydrogen-rich syngas production from chemical looping gasification of biomass char with $CaMn_{1-x}Fe_xO_3$. *Energy Fuels, 32*, 9541–9550.

Liu, G., Liao, Y., Wu, Y., & Ma, X., (2018). Application of calcium ferrites as oxygen carriers for microalgae chemical looping gasification. *Energy Convers. Manag., 160*, 262–272.

Liu, L., Huang, Y., & Liu, C., (2016). Prediction of rice husk gasification on fluidized bed gasifier based on aspen plus. *BioResources, 11*, 2744–2755.

Lopez, G., Alvarez, J., Amutio, M., Arregi, A., Bilbao, J., & Olazar, M., (2016). Assessment of steam gasification kinetics of the char from lignocellulosic biomass in a conical spouted bed reactor. *Energy, 107*, 493–501.

Lu, Q., Zhou, M., Li, W., Wang, X., Cui, M., & Yang, Y., (2018). Catalytic fast pyrolysis of biomass with noble metal-like catalysts to produce high-grade bio-oil: Analytical Py-GC/MS study. *Catal. Today, 302*, 169 179.

Lu, Y., Li, S., Guo, L., & Zhang, X., (2010). Hydrogen production by biomass gasification in supercritical water over $Ni/\gamma Al_2O_3$ and $Ni/CeO_2\text{-}\gamma Al_2O_3$ catalysts. *Int. J. Hydrogen Energy, 35*, 7161–7168.

Luo, M., Yi, Y., Wang, S., Wang, Z., Du, M., Pan, J., & Wang, Q., (2018). Review of hydrogen production using chemical-looping technology. *Renew. Sust. Energy Rev., 81*, 3186–3214.

Luo, S., Xiao, B., Guo, X., Hu, Z., Liu, S., & He, M., (2009). Hydrogen-rich gas from catalytic steam gasification of biomass in a fixed bed reactor: Influence of particle size on gasification performance. *Int. J. Hydrogen Energy, 34*, 1260–1264.

Lysikov, A., Derevschikov, V., & Okunev, A., (2015). Sorption-enhanced reforming of bioethanol in dual fixed bed reactor for continuous hydrogen production. *Int. J. Hydrogen Energy, 40*, 14436–14444.

Mahinpey, N., & Gomez, A., (2016). Review of gasification fundamentals and new findings: Reactors, feedstock, and kinetic studies. *Chem. Eng. Sci., 148*, 14–31.

Mayerhofer, M., Mitsakis, P., Meng, X., De Jong, W., Spliethoff, H., & Gaderer, M., (2012). Influence of pressure, temperature and steam on tar and gas in allothermal fluidized bed gasification. *Fuel, 99*, 204–209.

Mehrani, R., Barati, M., Tavasoli, A., & Karimi, A., (2015). Hydrogen production via supercritical water gasification of bagasse using $Ni\text{-}Cu/\gamma\text{-}Al_2O_3$ nano-catalysts. *Environ. Technol., 36*, 1265–1272.

Meshksar, M., Rahimpour, M., Daneshmand-Jahromi, S., & Hafizi, A., (2018). Synthesis and application of cerium-incorporated SBA-16 supported Ni-based oxygen carrier in cyclic chemical looping steam methane reforming. *Catalysts, 8*, 18.

Michel, R., Rapagnà, S., Di Marcello, M., Burg, P., Matt, M., Courson, C., & Gruber, R., (2011). Catalytic steam gasification of *Miscanthus X giganteus* in fluidized bed reactor on olivine based catalysts. *Fuel Process. Technol., 92*, 1169–1177.

Min, Z., Asadullah, M., Yimsiri, P., Zhang, S., Wu, H., & Li, C. Z., (2011). Catalytic reforming of tar during gasification. Part I. Steam reforming of biomass tar using ilmenite as a catalyst. *Fuel, 90*, 1847–1854.

Moghtaderi, B., Zanganeh, J., Shah, K., & Wu, H., (2012). Application of concrete and demolition waste as CO_2 sorbent in chemical looping gasification of biomass. *Energy Fuels, 26*, 2046–2057.

Mohammed, M. A. A., Salmiaton, A., Wan, A. W. A. K. G., Mohamad, A. M. S., Omar, R., Taufiq-Yap, Y. H., & Fakhru'l-Razi, A., (2012). Catalytic gasification of empty fruit bunch for enhanced production of hydrogen rich fuel gas. *Pertanika J. Sci. & Technol., 20*, 139–149.

Mohammed, M. A. A., Salmiaton, A., Wan, A. W. A. K. G., Mohammad, A. M. S., & Fakhru'l-Razi, A., (2011). Air gasification of empty fruit bunch for hydrogen-rich gas production in a fluidized-bed reactor. *Energy Convers. Manag., 52*, 1555–1561.

Mohanty, P., Pant, K. K., & Mittal, R., (2015). Hydrogen generation from biomass materials: Challenges and opportunities: Hydrogen generation from biomass materials. *WIREs Energy Environ., 4*, 139–155.

Molino, A., Chianese, S., & Musmarra, D., (2016). Biomass gasification technology: The state-of-the-art overview. *J. Energy Chem., 25*, 10–25.

Momirlan, M., & Veziroglu, T., (2005). The properties of hydrogen as fuel tomorrow in sustainable energy system for a cleaner planet. *Int. J. Hydrogen Energy, 30*, 795–802.

Nanda, S., & Berruti, F., (2021). A technical review of bioenergy and resource recovery from municipal solid waste. *J. Hazard. Mater., 403*, 123970.

Nanda, S., Dalai, A. K., Berruti, F., & Kozinski, J. A., (2016). Biochar as an exceptional bioresource for energy, agronomy, carbon sequestration, activated carbon and specialty materials. *Waste Biomass Valor., 7*, 201–235.

Nanda, S., Rana, R., Zheng, Y., Kozinski, J. A., & Dalai, A. K., (2017). Insights on pathways for hydrogen generation from ethanol. *Sustain. Energy Fuels, 1*, 1232–1245.

Newalkar, G., Iisa, K., D'Amico, A. D., Sievers, C., & Agrawal, P., (2014). Effect of temperature, pressure, and residence time on pyrolysis of pine in an entrained flow reactor. *Energy Fuels, 28*, 5144–5157.

Ni, Y., Wang, C., Chen, Y., Cai, X., Dou, B., Chen, H., Xu, Y., et al., (2017). High purity hydrogen production from sorption enhanced chemical looping glycerol reforming: Application of NiO-based oxygen transfer materials and potassium promoted Li_2ZrO_3 as CO_2 sorbent. *Appl. Ther. Eng., 124*, 454–465.

Okolie, J. A., Nanda, S., Dalai, A. K., & Kozinski, J. A., (2021). Chemistry and specialty industrial applications of lignocellulosic biomass. *Waste Biomass Valor, 12*, 2145–2169.

Okolie, J. A., Rana, R., Nanda, S., Dalai, A. K., & Kozinski, J. A., (2019). Supercritical water gasification of biomass: A state-of-the-art review of process parameters, reaction mechanisms and catalysis. *Sustain. Energy Fuels, 3*, 578–598.

Ortiz, M., Abad, A., De Diego, L. F., García-Labiano, F., Gayán, P., & Adánez, J., (2011). Optimization of hydrogen production by chemical-looping auto-thermal reforming working with Ni-based oxygen-carriers. *Int. J. Hydrogen Energy, 36*, 9663–9672.

Osada, M., Yamaguchi, A., Hiyoshi, N., Sato, O., & Shirai, M., (2012). Gasification of sugarcane bagasse over supported ruthenium catalysts in supercritical water. *Energy Fuels, 26*, 3179–3186.

Palancar, M. C., Serrano, M., & Aragón, J. M., (2009). Testing the technological feasibility of FLUMOV as gasifier. *Powder Technol., 194*, 42–50.

Palma, C. F., (2013). Modeling of tar formation and evolution for biomass gasification: A review. *Appl. Energy, 111*, 129–141.

Parthasarathy, P., & Narayanan, K. S., (2014). Hydrogen production from steam gasification of biomass: Influence of process parameters on hydrogen yield: A review. *Renew. Energy, 66*, 570–579.

Parthasarathy, P., & Sheeba, K. N., (2015). Combined slow pyrolysis and steam gasification of biomass for hydrogen generation-a review: Combined slow pyrolysis and steam gasification of biomass. *Int. J. Energy Res., 39*, 147–164.

Peng, W. X., Wang, L. S., Mirzaee, M., Ahmadi, H., Esfahani, M. J., & Fremaux, S., (2017). Hydrogen and syngas production by catalytic biomass gasification. *Energy Convers. Manag., 135*, 270–273.

Pfeifer, C., & Hofbauer, H., (2008). Development of catalytic tar decomposition downstream from a dual fluidized bed biomass steam gasifier. *Powder Technol., 180*, 9–16.

Protasova, L., & Snijkers, F., (2016). Recent developments in oxygen carrier materials for hydrogen production via chemical looping processes. *Fuel, 181*, 75–93.

Qiu, Q., Bo, Z., & Weizhong, L., (2018). Green energy based thermochemical and photochemical hydrogen production. *Int. J. Electrochem. Sci.*, 6484–6502.

Radwan, A. M., (2012). An overview on gasification of biomass for production of hydrogen rich gas. *Der Chemica Sinica, 3*(2), 323–335.

Rapagna, S., Jand, N., & Foscolo, P. U., (1998). Catalytic gasification of biomass to produce hydrogen rich gas. *Int. J. Hydrogen Energy, 23*, 551–557.

Rapagnà, S., Virginie, M., Gallucci, K., Courson, C., Di Marcello, M., Kiennemann, A., & Foscolo, P. U., (2011). Fe/olivine catalyst for biomass steam gasification: Preparation, characterization and testing at real process conditions. *Catal. Today, 176*, 163–168.

Roses, L., Campanari, S., & Manzolini, G., (2013). Computational fluid dynamics (CFD) analysis of membrane reactors: Simulation of a palladium-based membrane reactor in fuel cell micro-cogenerator system. In: *Handbook of Membrane Reactors* (pp. 496–531). Elsevier.

Rossetti, I., Compagnoni, M., Finocchio, E., Ramis, G., Di Michele, A., Zucchini, A., & Dzwigaj, S., (2016). Syngas production via steam reforming of bioethanol over Ni-BEA catalysts: A BTL strategy. *Int. J. Hydrogen Energy, 41*, 16878–16889.

Rostrup-Nielsen, J. R., Sehested, J., & Nørskov, J. K., (2002). Hydrogen and synthesis gas by steam- and CO_2 reforming. *Adv. Catal., 47*, 65–139.

Russell, S. H., Turrion-Gomez, J. L., Meredith, W., Langston, P., & Snape, C. E., (2017). Increased charcoal yield and production of lighter oils from the slow pyrolysis of biomass. *J. Anal. Appl. Pyrolysis, 124*, 536–541.

Salam, M. A., Ahmed, K., Akter, N., Hossain, T., & Abdullah, B., (2018). A review of hydrogen production via biomass gasification and its prospect in Bangladesh. *Int. J. Hydrogen Energy, 43*, 14944–14973.

Salkuyeh, Y. K., Saville, B. A., & MacLean, H. L., (2017). Techno-economic analysis and life cycle assessment of hydrogen production from natural gas using current and emerging technologies. *Int. J. Hydrogen Energy, 42*, 18894–18909.

Sandeep, K., & Dasappa, S., (2014). Oxy-steam gasification of biomass for hydrogen rich syngas production using downdraft reactor configuration: Oxy-steam gasification of biomass for hydrogen rich syngas production. *Int. J. Energy Res., 38*, 174–188.

Sansaniwal, S. K., Pal, K., Rosen, M. A., & Tyagi, S. K., (2017a). Recent advances in the development of biomass gasification technology: A comprehensive review. *Renew. Sust. Energy Rev., 72*, 363–384.

Sansaniwal, S. K., Rosen, M. A., & Tyagi, S. K., (2017b). Global challenges in the sustainable development of biomass gasification: An overview. *Renew. Sust. Energy Rev., 80*, 23–43.

Sharma, S., & Sheth, P. N., (2016). Air-steam biomass gasification: Experiments, modeling and simulation. *Energy Convers. Manag., 110*, 307–318.

Shayan, E., Zare, V., & Mirzaee, I., (2018). Hydrogen production from biomass gasification; a theoretical comparison of using different gasification agents. *Energy Convers. Manag., 159*, 30–41.

Shen, Y., Zhao, P., Shao, Q., Takahashi, F., & Yoshikawa, K., (2015). In situ catalytic conversion of tar using rice husk char/ash supported nickel-iron catalysts for biomass pyrolytic gasification combined with the mixing-simulation in fluidized-bed gasifier. *Appl. Energy, 160*, 808–819.

Shokrollahi, Y. M., Radfarnia, H. R., & Iliuta, M. C., (2017). Sustainable production of high-purity hydrogen by sorption enhanced steam reforming of glycerol over CeO_2-promoted $Ca_9Al_6O_{18}$–CaO/NiO bifunctional material. *ACS Sustain. Chem. Eng., 5*, 9774–9786.

Sikarwar, V. S., Zhao, M., Clough, P., Yao, J., Zhong, X., Memon, M. Z., Shah, N., Anthony, E. J., & Fennell, P. S., (2016). An overview of advances in biomass gasification. *Energy Environ. Sci., 9*, 2939–2977.

Skoulou, V., Swiderski, A., Yang, W., & Zabaniotou, A., (2009). Process characteristics and products of olive kernel high temperature steam gasification (HTSG). *Bioresour. Technol., 100*, 2444–2451.

Spragg, J., Mahmud, T., & Dupont, V., (2018). Hydrogen production from bio-oil: A thermodynamic analysis of sorption-enhanced chemical looping steam reforming. *Int. J. Hydrogen Energy, 43*, 22032–22045.

Srinakruang, J., Sato, K., Vitidsant, T., & Fujimoto, K., (2005). A highly efficient catalyst for tar gasification with steam. *Catal. Comm., 6*, 437–440.

Striūgas, N., Zakarauskas, K., Stravinskas, G., & Grigaitienė, V., (2012). Comparison of steam reforming and partial oxidation of biomass pyrolysis tars over activated carbon derived from waste tire. *Catal. Today, 196*, 67–74.

Sun, L., Zhang, X., Chen, L., Zhao, B., Yang, S., & Xie, X., (2016). Comparision of catalytic fast pyrolysis of biomass to aromatic hydrocarbons over ZSM-5 and Fe/ZSM-5 catalysts. *J. Anal. Appl. Pyrolysis, 121*, 342–346.

Sutton, D., Kelleher, B., & Ross, J. R. H., (2001). Review of literature on catalysts for biomass gasification. *Fuel Process. Technol., 73*, 155–173.

Umeki, K., Yamamoto, K., Namioka, T., & Yoshikawa, K., (2010). High temperature steam-only gasification of woody biomass. *Appl. Energy, 87*, 791–798.

Van, D. V. M., Baeyens, J., Brems, A., Janssens, B., & Dewil, R., (2010). Fundamentals, kinetics and endothermicity of the biomass pyrolysis reaction. *Renew. Energy, 35*, 232–242.

Wang, C., Dou, B., Jiang, B., Song, Y., Du, B., Zhang, C., Wang, K., Chen, H., & Xu, Y., (2015). Sorption-enhanced steam reforming of glycerol on Ni-based multifunctional catalysts. *Int. J. Hydrogen Energy, 40*, 7037–7044.

Wang, D., Yuan, W., & Ji, W., (2011). Char and char-supported nickel catalysts for secondary syngas cleanup and conditioning. *Appl. Energy, 88*, 1656–1663.

Wang, G., Xu, S., Wang, C., & Zhang, J., (2017). Biomass gasification and hot gas upgrading in a decoupled dual-loop gasifier. *Energy Fuels, 31*, 8181–8192.

Wang, K., Yu, Q., Qin, Q., Hou, L., & Duan, W., (2016). Thermodynamic analysis of syngas generation from biomass using chemical looping gasification method. *Int. J. Hydrogen Energy, 41*, 10346–10353.

Wang, L., Weller, C. L., Jones, D. D., & Hanna, M. A., (2008). Contemporary issues in thermal gasification of biomass and its application to electricity and fuel production. *Biomass Bioenergy, 32*, 573–581.

Wang, S., Dai, G., Yang, H., & Luo, Z., (2017). Lignocellulosic biomass pyrolysis mechanism: A state-of-the-art review. *Prog. Energy Combust. Sci., 62*, 33–86.

Wang, Z., He, T., Qin, J., Wu, J., Li, J., Zi, Z., Liu, G., et al., (2015). Gasification of biomass with oxygen-enriched air in a pilot scale two-stage gasifier. *Fuel, 150*, 386–393.

Weerachanchai, P., Horio, M., & Tangsathitkulchai, C., (2009). Effects of gasifying conditions and bed materials on fluidized bed steam gasification of wood biomass. *Bioresour. Technol., 100*, 1419–1427.

Wei, G., He, F., Zhao, Z., Huang, Z., Zheng, A., Zhao, K., & Li, H., (2015). Performance of Fe-Ni bimetallic oxygen carriers for chemical looping gasification of biomass in a 10 kWth interconnected circulating fluidized bed reactor. *Int. J. Hydrogen Energy, 40*, 16021–16032.

Wei, L., Xu, S., Zhang, L., Liu, C., Zhu, H., & Liu, S., (2007). Steam gasification of biomass for hydrogen-rich gas in a free-fall reactor. *Int. J. Hydrogen Energy, 32*, 24–31.

Wilk, V., & Hofbauer, H., (2013). Conversion of mixed plastic wastes in a dual fluidized bed steam gasifier. *Fuel, 107*, 787–799.

Wolfesberger, U., Aigner, I., & Hofbauer, H., (2009). Tar content and composition in producer gas of fluidized bed gasification of wood-Influence of temperature and pressure. *Environ. Prog. Sustain. Energy, 28*, 372–379.

Wright, M. M., Daugaard, D. E., Satrio, J. A., & Brown, R. C., (2010). Techno-economic analysis of biomass fast pyrolysis to transportation fuels. *Fuel, 89*, S2–S10.

Wu, Y., Liao, Y., Liu, G., & Ma, X., (2018). Syngas production by chemical looping gasification of biomass with steam and CaO additive. *Int. J. Hydrogen Energy, 43*, 19375–19383.

Xie, H., Yu, Q., Lu, H., Zhang, Y., Zhang, J., & Qin, Q., (2017). Thermodynamic study for hydrogen production from bio-oil via sorption-enhanced steam reforming: Comparison with conventional steam reforming. *Int. J. Hydrogen Energy, 42*, 28718–28731.

Xu, C. C., Donald, J., Byambajav, E., & Ohtsuka, Y., (2010). Recent advances in catalysts for hot-gas removal of tar and NH_3 from biomass gasification. *Fuel, 89*, 1784–1795.

Xu, C., Chen, S., Soomro, A., Sun, Z., & Xiang, W., (2018). Hydrogen rich syngas production from biomass gasification using synthesized Fe/CaO active catalysts. *J. Energy Inst., 91*, 805–816.

Xue, Z., Shen, Y., Zhu, S., Li, P., Zeng, Y., Xi, Z., & Cai, Y., (2017). Autothermal reforming of ethyl acetate for hydrogen production over $Ni_3La_7O_y/Al_2O_3$ catalyst. *Energy Convers. Manag., 146*, 34–42.

Yakaboylu, O., Albrecht, I., Harinck, J., Smit, K. G., Tsalidis, G. A., Di Marcello, M., Anastasakis, K., & De Jong, W., (2018). Supercritical water gasification of biomass in fluidized bed: First results and experiences obtained from TU Delft/Gensos semi-pilot scale setup. *Biomass Bioenergy, 111*, 330–342.

Yang, X., Wang, Y., Li, M., Sun, B., Li, Y., & Wang, Y., (2016). Enhanced hydrogen production by steam reforming of acetic acid over a Ni catalyst supported on mesoporous MgO. *Energy Fuels, 30*, 2198–2203.

Yanik, J., Ebale, S., Kruse, A., Saglam, M., & Yuksel, M., (2007). Biomass gasification in supercritical water: Part 1. Effect of the nature of biomass. *Fuel, 86*, 2410–2415.

Yao, J., Kraussler, M., Benedikt, F., & Hofbauer, H., (2017). Techno-economic assessment of hydrogen production based on dual fluidized bed biomass steam gasification, biogas steam reforming, and alkaline water electrolysis processes. *Energy Convers. Manag., 145*, 278–292.

Yuan, T., Tahmasebi, A., & Yu, J., (2015). Comparative study on pyrolysis of lignocellulosic and algal biomass using a thermogravimetric and a fixed-bed reactor. *Bioresour. Technol., 175*, 333–341.

Zamzuri, N. H., Mat, R., Saidina, A. N. A., & Talebian-Kiakalaieh, A., (2017). Hydrogen production from catalytic steam reforming of glycerol over various supported nickel catalysts. *Int. J. Hydrogen Energy, 42*, 9087–9098.

Zeng, J., Xiao, R., Zhang, H., Chen, X., Zeng, D., & Ma, Z., (2017). Syngas production via biomass self-moisture chemical looping gasification. *Biomass Bioenergy, 104*, 1–7.

Zhang, B., Zhang, L., Yang, Z., Yan, Y., Pu, G., & Guo, M., (2015). Hydrogen-rich gas production from wet biomass steam gasification with CaO/MgO. *Int. J. Hydrogen Energy, 40*, 8816–8823.

Zhang, Z., & Pang, S., (2017). Experimental investigation of biomass devolatilization in steam gasification in a dual fluidized bed gasifier. *Fuel, 188*, 628–635.

Zhao, Z., Uddi, M., Tsvetkov, N., Yildiz, B., & Ghoniem, A. F., (2016). redox kinetics study of fuel reduced ceria for chemical-looping water splitting. *J. Phys. Chem. C, 120*, 16271–16289.

Zhou, J., Masutani, S. M., Ishimura, D. M., Turn, S. Q., & Kinoshita, C. M., (2000). Release of fuel-bound nitrogen during biomass gasification. *Ind. Eng. Chem. Res., 39*, 626–634.

Zhu, X., Zhang, M., Li, K., Wei, Y., Zheng, Y., Hu, J., & Wang, H., (2018). Chemical-looping water splitting over ceria-modified iron oxide: Performance evolution and element migration during redox cycling. *Chem. Eng. Sci., 179*, 92–103.

CHAPTER 3

Hydrogen Production by Catalytic Reforming Process

CHANDRAMANI RAI and PRABU VAIRAKANNU

Department of Chemical Engineering, Indian Institute of Technology Guwahati, Guwahati, Assam, India
E-mail: v.prabu@iitg.ac.in (Prabu Vairakannu)

ABSTRACT

In recent times, clean energy production is one of the major challenges that the world is facing in energy sectors as most of the energy need is supplied either by coal or by petroleum or natural gas. The utilization of fossil fuels directly emits greenhouses gases such as CO, CO_2, and NO_x impurities, which could lead to global warming. Hence, by the pre-combustion capture method, these gases need to be captured and fossil fuels should be converted into clean energy such as hydrogen. Natural gas and various chemicals from coal and petroleum feedstock can be converted into hydrogen in several ways. One of the routes is the steam reforming technology by catalytic conversion. This chapter makes a comprehensive review of the catalytic reforming technologies for hydrogen production. Various feedstock, conversion processes, hydrogen selectivity, and the choice of catalyst are discussed.

3.1 INTRODUCTION

In most recent years, developing and developed countries are facing many crises in the energy sector, particularly due to fossil fuel depletion and their sources. The burning of non-sustainable power sources, specific petroleum products and their derivative provides most of the current energy needs. These fuels produced enormous amounts of greenhouse gases, particularly CO_2, CO, and other harmful gases to the atmosphere (Amin et al., 2011). The gradual decrease of fossil fuel and the emission of greenhouse gases make compulsive use of alternate sources of energy (Noh et al., 2019). Hydrogen is one of the important raw materials in chemical industries and energy sectors because of its availability, sustainability, and eco-friendly nature (Sarangi and Nanda, 2020).

Hydrogen can be considered as one of the clean fuels of the future. It can be produced through different hydrocarbon feedstocks such as natural gas, coal, biomass, alcohol, and water (Bizkarra et al., 2019). There are various processes available to produce hydrogen from different feedstock by reforming, gasification, cracking, electrolysis, etc. (Nanda et al., 2017; Nguyen et al., 2019). Reforming is the prominent technology available to produce hydrogen. It is based on the endothermic or exothermic transformation of the organic precursor with $H_2O/CO_2/O_2$ into syngas (a mixture of CO and H_2). The most frequently used feedstocks for the production of hydrogen are methane and other light hydrocarbons. There are several processes, which produce hydrogen via reforming technique, such as steam reforming (SR), autothermal reforming (ATR), partial oxidation (POX), dry reforming of methane (DMR), combined reforming of methane (CMR), reforming with the membrane, tri-reforming of methane (TMR) and chemical looping reforming (CLR).

Steam gasification and partial oxidation methods are widely used in industry to produce hydrogen, and others are novel processes developed with the aim of maximization of hydrogen production and minimization of environmental impact. Other various alternative processes such as pyrolysis, CO_2 gasification, cracking, and water splitting are used to produce hydrogen from various feedstocks (Vozniuk et al., 2019). The ratio of H_2:CO in the syngas depends on the reaction from which it is generated, such as dry reforming of methane (H_2:CO = 1), partial oxidation of methane (H_2:CO = 2) and steam reforming of methane (H_2:CO = 3).

Hydrogen is used as a raw material in various industries including the fertilizer sector, the petroleum sector and the chemical industry. In fertilizer sectors, hydrogen is mostly used in the production of ammonia and methanol (approximately 50% of total production). The petroleum sectors use approximately 30% of total production, mainly for hydrotreating and hydrocracking of fuel. The other use of hydrogen is in food industries (i.e., sorbitol and fat processing), metal industries (ore reduction) and semiconductor industries. Pure hydrogen is mainly used for the processing of unsaturated hydrocarbons and aromatics (Spragg et al., 2018). It can play an important role in the clean energy fuel field including transport, heating, and energy storage (Ball and Weeda, 2015; Chen et al., 2020). Hydrogen can help to produce carbon-free energy. It is used in fuel cells for automotive purposes (Marban and Valdes-Solis, 2007; Dodds and McDowall, 2013).

3.2 CATALYTIC STEAM REFORMING

Catalytic steam reforming is the most common technique used for hydrogen production. This is widely used because of their high-performance efficiency, low implication cost and installation simplicity as compared to other technologies. This technology produces approximately 50% of hydrogen of all needs by using different types of feedstocks such as methane, ethane, dodecane, natural gases, naphtha, alcohol, etc. The primary product of steam reforming is CO, H_2, CO_2 and CH_4. The yield of the specified product depends upon various factors such as temperature, pressure, reformer type, catalyst, and other factors (Tuna et al., 2018). Various types of catalysts used in this process are primary noble metals (e.g., Rh, Ru, Pd and Pt) and transition metals (e.g., Ni, Co, Fe, and Cu) with different supports (e.g., Al_2O_3 and SiO_2) (Yadav and Vaidya, 2019). The steam reforming and water-gas shift reactions are shown below.

$$CH_4 + H_2O \rightleftharpoons CO + 3H_2 \tag{3.1}$$

$$CO + H_2O \rightleftharpoons CO_2 + H_2 \tag{3.2}$$

Steam reforming reaction is endothermic and reversible. According to Le-Chatelier's principle, to achieve high conversion, temperature, and steam-to-methane ratio should be high, and the operating pressure must be low. These processes are performed at around 800°C in a fixed bed and fluidized bed reactors (Yoo et al., 2017). Other feedstocks such as alcohol and acids can be reformed into syngas. Acetic acid is reformed as per the following reaction (Kumar et al., 2019).

$$CH_3COOH + 2H_2O \rightleftharpoons 2CO_2 + 4H_2 \tag{3.3}$$

Ethanol and methanol are transformed into CH_4, CO, and H_2 as per the following reactions (Wang et al., 2019, Hwang et al., 2019).

$$2C_2H_5OH \rightleftharpoons 3CH_4 + CO_2 \tag{3.4}$$

$$C_2H_5OH \rightleftharpoons 3CH_4 + CO + H_2 \tag{3.5}$$

The steam reforming reactions of acetic acid, ethanol, and methanol are reversible and endothermic and are performed in a fixed bed or fluidized bed, depending upon the feed quality, quantity, conversion, and yield of a certain product. Figure 3.1 shows the catalytic steam reforming process. Firstly, catalysts are loaded in the middle section of the reactor. Then the raw materials (e.g., natural gas, ethanol, methanol, acetic acid) and steam are fed into the reactor in a specific steam-to-fuel ratio. The operating temperature in the range of 600–800°C needs to be maintained (Wang et al., 2019; Hwang et al., 2019).

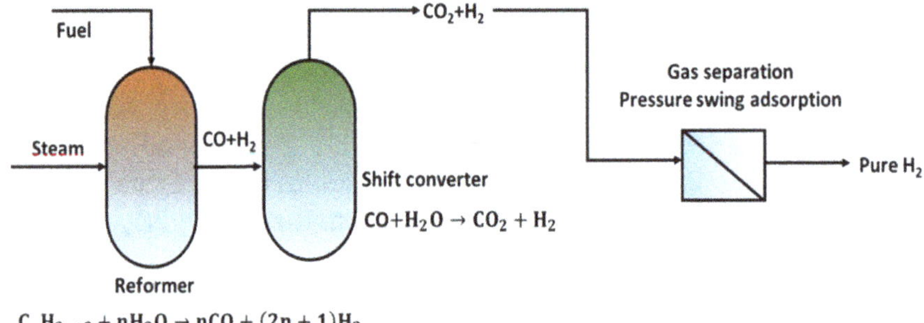

FIGURE 3.1 Schematic representation of the steam reforming process.

Table 3.1 shows the conversion of various feedstock and yield of hydrogen. It can be seen that acetic acid, dodecane, ethanol, methanol, methane, ethane, and tar content are the potential source for the production of hydrogen through the steam reforming process. One can see the selectivity of hydrogen is found highest under different combinations of feedstocks and catalysts. In the case of acetic acid reforming, nickel, and cobalt showed 90% H_2 selectivity and 100% conversion of the feedstock (Hu and Lu, 2010), whereas dodecane revealed H_2 selectivity of 22% with nickel-platinum catalyst (Li et al., 2020). Ethanol exhibited a better steam reforming tendency using a copper-cesium catalyst with more than 90% conversion and H_2 selectivity occurred even at low temperatures of 300°C. Methanol had shown only 65% H_2 selectivity using platinum-Ln_2O_3-Al_2O_3 catalysts. Methane and ethane steam reforming occur effectively (above 80% conversion) at 600°C with nickel embedded alumina as the catalysts. Further, tar generated from biomass exhibited a good reforming tendency using nickel-lanthanum oxide catalysts. Thus, Ni-based catalyst is suitable for most of the feedstocks for the steam reforming process.

TABLE 3.1 Catalytic Steam Reforming of Various Feedstocks

Precursor	Catalyst	Temperature (°C)	Conversion (%)	H_2 selectivity	Reactor	Steam/carbon ratio	Products	References
Acetic acid	• Rh(0.5%)/Al_2O_3 • Ru(1%)/Al_2O_3 • Pd(1%)/Al_2O_3 • Pt(1%)/Al_2O_3 • Ni(20%)/Al_2O_3 • Co(20%)/Al_2O_3 • Fe(20%)/Al_2O_3 • Cu(20%)/Al_2O_3 • Ni(10%)/Al_2O_3 • Ni(10%)/Al_2O_3	• 800 • 800 • 800 • 800 • 823 • 823 • 823 • 823 • 600 • 600	• 97 • 38 • 37 • 29 • 100 • 100 • 45 • 50 • 100 • 100	• 96 • 90 • 91 • 75 • 92 • 90 • 34 • 23 • 70 (yield) • 30 (yield)	Fixed bed	-	CO, CO_2, H_2	• Basagiannis and Verykios (2007) • Basagiannis and Verykios (2007) • Basagiannis and Verykios (2007) • Basagiannis and Verykios (2007) • Hu and Lu (2010) • Hu and Lu (2010) • Hu and Lu (2010) • Hu and Lu (2010) • Zhang et al. (2019) • Zhang et al. (2019)
Dodecane	• Ni/Al_2O_3 • NiPt0.5/Al_2O_3 • NiPt1.5/Al_2O_3 • NiPt/Al_2O_3	• 700 • 700 • 700 • 700	• 19 • 65 • 77 • 93	• 3.8 • 15.5 • 16.8 • 21.7	Fixed bed		CO, H_2	Li et al. (2020)

TABLE 3.1 *(Continued)*

Precursor	Catalyst	Temperature (°C)	Conversion (%)	H_2 selectivity	Reactor	Steam/carbon ratio	Products	References
Ethanol	• Co-Birnsite • Co-Todorokite • Ni/Co(0.9)Sm(0.1)O(2–8) • Ni-Cu-Cs/Activated carbon • Ni-Cu-Cs/LiAl_2O_3 • Cu(15%)-Cs(2%)/Al_2O_3 • Cu(15%)/Al_2O_3 • Cu(15%)-Cs(1%)/Al_2O_3 • Cu(15%)-Cs(3%)/Al_2O_3	• 773 • 773 • 550 • 300 • 300 • 300 • 300 • 300 • 300	• 94.6 • 87.9 • 100 • 98 • 95 • 94 • 84 • 88 • 90	• 69.8 • 74.9 • 60 • 26 • 56 • 97 • 86 • 90 • 92	Fixed bed quartz reactor	• 3 • 1.85 • 1.85	• CO, H_2 • CO_2, CH_4, H_2 • CO_2, CH_4, H_2 • H_2, CO_2 • CO, H_2, CO_2 • CO, H_2, CO_2 • CO, H_2, CO_2 • CO, H_2, CO_2	• Da Costa-Serra and Chica (2018) • Da Costa-Serra and Chica (2018) • Rodrigues et al. (2019) • Özkan et al. (2018) • Özkan et al. (2018) • Houteit et al. (2006) • Houteit et al. (2006) • Houteit et al. (2006) • Houteit et al. (2006)

TABLE 3.1 *(Continued)*

Precursor	Catalyst	Temperature (°C)	Conversion (%)	H_2 selectivity	Reactor	Steam/ carbon ratio	Products	References
Methanol	• Nb(1%)/Pd-Zr-Zn • Pd-Zr-Zn • Pt(15%)/CeO_2 • Pt(15%)/ Ln_2O_3(15%)CeO_2 • Pt(15%)/ Ln_2O_3(15%)Al_2O_3 • Pt(15%)/Ln_2O_3 (15%)ZrO_2 • Pt(15%)/Ln_2O_3 (30%)CeO_2 • Pt(15%)/ Ln_2O_3(15%)Al_2O_3	• 300 • 300 • 350 • 350 • 350 • 350 • 350 • 350	• 81.2 • 85 • 100 • 99.9 • 99.4 • 99.8 • 98.9 • 95.9	- - • 61.1 • 64.7 • 64.3 • 3.4 • 65.2 • 65.0	Fixed bed	• 1.4 • 1.4 • 1.4 • 1.4 • 1.4 • 1.4	• H_2, CO, CH_4 • H_2, CO, CH_4 • H_2, CO, CO_2 • H_2, CO, CO_2 • H_2, CO, CO_2 • H_2, CO, CO_2 • H_2, CO, CO_2 • H_2, CO, CO_2	• Cai et al. (2018) • Cai et al. (2018) • Shanmugam et al. (2020) • Shanmugam et al. (2020) • Shanmugam et al. (2020) • Shanmugam et al. (2020) • Shanmugam et al. (2020) • Shanmugam et al. (2020)
Methane	• 20%Ni/SiO_2 • 20%Ni/Al_2O_3	• 500 • 500	• 23 • 24	• 42 • 41			CO_2, CO, H_2	Matsumura and Nakamori (2004)
Methane and ethane (4.6:0.4 ratio)	• Ni/Al_2O_3 (trimethylbenzene) • Ni/Al_2O_3	• 600 • 600	• 96 • 89	• 97 • 94			CO_2, CO, H_2	Bang et al. (2013)

TABLE 3.1 *(Continued)*

Precursor	Catalyst	Temperature (°C)	Conversion (%)	H_2 selectivity	Reactor	Steam/carbon ratio	Products	References
Methane (Ni/Al atomic ratio 0.35)	• Ni-Al_2O_3 (Na-Es) Epoxide super critical • Ni-Al_2O_3 (Na-As) • Alkoxide supercritical	• 600 • 600	• 85 • 80	• 86 • 79			CO_2, CO, H_2	Bang et al. (2012)
Liquefied natural gas (92% CH_4 and 8% C_2H_6) (CH_4:C_2H_6 ratio = 4.6:0.4)	• NiNi/Al_2O_3 • NiCe/Al_2O_3 • NiCd/Al_2O_3 • NiY/Al_2O_3 • NiCs/Al_2O_3 • NiFe/Al_2O_3 • NiCo/Al_2O_3 • NiMg/Al_2O_3	600	• 77 • 87 • 85 • 83 • 80 • 77 • 78 • 71	• 138% (yield) • 140% (yield) • 141% (yield) • 140% (yield) • 139% (yield) • 128% (yield) • 124% (yield) • 121% (yield)	Fixed bed reactor	2	CO_2, CO, H_2	Seo et al. (2011)

TABLE 3.1 *(Continued)*

Precursor	Catalyst	Temperature (°C)	Conversion (%)	H_2 selectivity	Reactor	Steam/carbon ratio	Products	References
Gasified biomass tar (15 wt.% phenol, 50 wt.% toluene, 30 wt.% naphthalene and 5 wt.% pyrene)	Ni/dolomite La_2O_3 catalyst	750	89	81	Fixed bed	1	CO_2, CO, H_2	Tana et al. (2019)

3.3 DRY REFORMING OF METHANE

Dry reforming of methane produces syngas in an equimolar ratio of CO and H_2 using CO_2. This reaction is desirable, compared to others as it consumes greenhouse gases (both reactants such as CH_4 and CO_2) to produce the syngas (Evans et al., 2014; Siang et al., 2018). Figure 3.2 shows the schematic representation of the dry reforming process.

$$CH_4 + CO_2 \rightleftharpoons 2CO + 2H_2 \tag{3.6}$$

The following side reactions occur along with the dry reforming reaction. Reverse water-gas shift reaction:

$$CO_2 + H_2 \rightleftharpoons CO + H_2O \tag{3.7}$$

Cracking reaction:

$$CH_4 \rightleftharpoons 2H_2 + C \tag{3.8}$$

Disproportionation reaction:

$$2CO \rightleftharpoons CO_2 + C \tag{3.9}$$

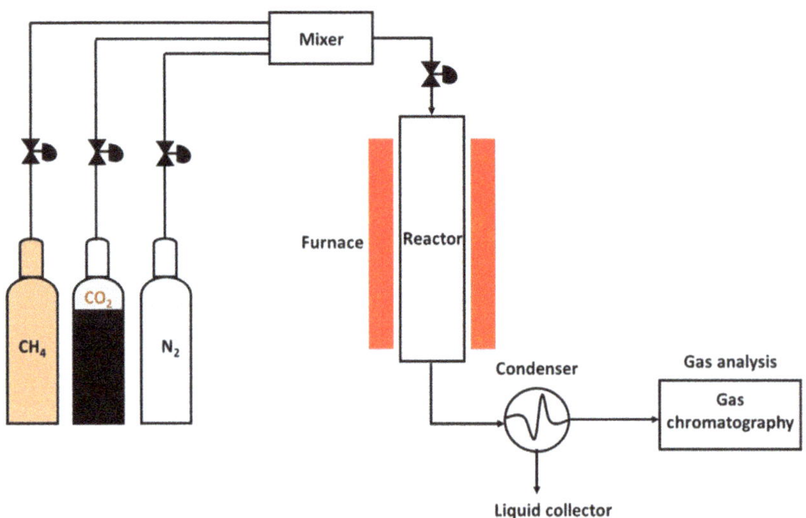

FIGURE 3.2 Schematic representation of dry reforming of methane.

Reverse water-gas shift and disproportionation reaction are suppressed towards the formation of CO, which deteriorates the catalyst activity (Abdullah et al., 2020). The selection of catalyst plays an important role in the dry reforming of methane. One of the biggest challenges in the case of this process is the preparation of catalyst in such a way that it restricts the formation of coke. There are various types of catalysts such as noble metals (e.g., Rh, Ru, Pd, and Pt) and transition metals (e.g., Ni, Co, Fe, and Cu) (Zhang et al., 2017; Abdulrasheed et al., 2020) for dry reforming of methane. This process is conducted using either a fixed bed or a fluidized bed reactor. As the operational cost of the fluidized bed is high, a fixed bed condition is preferred under the pilot-scale level. The quality of the gases produced during this process depends upon the types of reactors, catalysts, reactor temperature, pressure, and reactant ratio (CH_4 and CO_2). Furthermore, the catalyst preparation method plays an important role in the conversion and selectivity of the produced gas (Hambali et al., 2020). This reaction is also endothermic nature that requires the maintenance of high operating temperatures. The use of catalysts provides an alternate path to reduce the required temperature.

Table 3.2 shows the summary of the conversion of methane and selectivity of hydrogen using various catalysts. The operating temperature of the dry reforming of methane is in the range of 700–900°C. It can be seen that most of the studies used nickel with a combination of other catalysts. Among these studies, nickel combination with cesium-calcium impregnated mesoporous silica carbon (MSC) achieved a higher conversion (93%) of CH_4 and CO_2. Higher Ce/Ca ratio in the catalyst improved the dispersibility resulting in more active sites, whereas Co, Pt, and Ni alone as catalysts showed a low conversion in the range of 50–70%.

3.4 PARTIAL OXIDATION OF METHANE

Partial oxidation of methane becomes an important process for the production of hydrogen in recent years. As the nature of reaction during partial oxidation is exothermic, this makes the process less expensive (10–15% energy reduction) and low capital investment (25–30% reduction) as compared to the steam reforming process (Corbo and Migliardini, 2007). However, the exothermic nature of the reaction creates a hotspot on the catalyst surface leading to operational difficulty. The syngas produced during this process with an H_2/CO ratio of 2 would be suitable for Fischer-Tropsch synthesis of methanol (Pantaleo et al., 2016). High methane conversion and syngas yield are reported with a multi-second contact time of the operation (Osman et al., 2017). The preparation of an efficient and stable catalyst is the main challenging problem in the catalytic partial oxidation process.

TABLE 3.2 Conversion and Selectivity of Hydrogen by Dry Reforming of Methane

Precursor	Catalyst	Temperature (°C)	Conversion (%)	H_2/CO_2 ratio in the product gas	Products	References
Methane	Ni/MFI (Wetness impregnation method)	800	CH_4: 70 CO_2: 72	0.95	CO, H_2	Hambali et al. (2020)
	Ni/MFI (Physical method)		CH_4: 60 CO_2: 62	0.89		
	Ni/MFI (Double solvent)		CH_4: 55 CO_2: 53	0.79		
Methane	Ni-5Ca-Mesoporous silica carbon (MSC)	900	CH_4: 92 CO_2: 93	0.98	CO, H_2	Sun et al. (2020)
	Ni-1Ce-4Ca-MSC		CH_4: 94 CO_2: 94	0.97		
	Ni-2Ce-3Ca-MSC		CH_4: 93 CO_2: 95	0.95		
	Ni-3Ce-2Ca-MSC		CH_4: 96 CO_2: 95	0.92		
	Ni-4Ce-1Ca-MSC		CH_4: 95 CO_2: 96	0.97		
	Ni-5Ce-MSC		CH_4: 90 CO_2: 93	0.99		

TABLE 3.2 *(Continued)*

Precursor	Catalyst	Temperature (°C)	Conversion (%)	H_2/CO_2 ratio in the product gas	Products	References
CH_4:CO_2 (1:1)	4% Ni/Al_2O_3	700	CH_4: 60 CO_2: 56			García-Diéguez et al. (2010)
CH_4:CO_2 (1:1)	5% Co/Al_2O_3	700	CH_4: 52 CO_2: 52			Özkara-Aydınoğlu et al. (2010)
$CH_{4:}CO_2$ (1:1)	4% Pt-4%Ni/Al_2O_3	700	CH_4: 69 CO_2: 76			Özkara-Aydınoğlu et al. (2010)
CH_4:CO_2 (1:1)	5% Ni/10%Co/Al_2O_3	700	CH_4: 67 CO_2: 71			Siang et al. (2017)

TABLE 3.3 Partial Oxidation of Methane for Hydrogen Production

Precursor	Catalyst	Temperature (°C)	Conversion (%)	Selectivity (%)	Reactors	Steam-to-carbon ratio product	References
10 vol% CH_4, 5 vol% O_2, 85 vol% Ar	25% SPHERE	750	87	88	Quartz tube fixed bed reactor	CO_2, H_2	Ma et al. (2019)
	35% SPHERE	750	74	70			
	25% FIBRE	750	14	81			
	35% FIBRE	750	83	95			
	45% FIBRE	750	81	91			
10% CH_4, 5% O_2, 80% Ar	Thermodynamic study	705	96.5			CO_2, CO, H_2, CH_4, H_2O, Ar	Osman (2020)
		800	98.6				
20% CH_4, 10% O_2, 65% Ar		705	94.3				
		800	97.6	0.3			

Partial oxidation of methane using a noble catalyst is quite efficient and carbon resistant, but the cost of these noble metals is high. The transition metal, nickel is a good alternative to noble metals in terms of activity and low cost for this process to catalyze methane, but the main drawback of these elements is carbon formation, which deteriorates the activity of the catalyst. In the catalytic oxidation process, firstly, catalyst is fed into the reactors. Then, methane (2 vol%) and oxygen (1 vol%) are supplied using a pump. Catalyst performance is measured in terms of conversion of methane and yield of hydrogen. Table 3.3 shows the summary of partial oxidation of methane for hydrogen production. A specific study on the NiO-based catalyst in fibrous form had shown a higher hydrogen selectivity of 95%, compared to conventional monolithic catalysts (70–80% selectivity). The fibrous catalyst exhibits a higher mass transfer rate and higher resistance to carbon deposition.

3.5 AUTOTHERMAL REFORMING OF METHANE

Auto thermal reforming is an important process to produce synthesis gas with a low H_2/CO ratio of 1.5. Figure 3.3 shows the schematic view of the autothermal reforming of methane. The syngas obtained during this process is used to produce methanol and higher molecular weight hydrocarbons. This process is a combination of catalytic partial oxidation and steam reforming. The optimal temperature is maintained by using the heat obtained from the partial oxidation of hydrocarbons using a stoichiometric amount of air. One of the basic auto thermal reforming processes of methane is (Gallucci et al., 2010).

$$CH_4 + 0.5\ O_2 \rightleftharpoons CO + 2H_2O \tag{3.10}$$

Steam produced in the partial oxidation reaction (Eq. 3.10) is used for the reforming of methane. The various types of feedstocks such as methane, ethanol, acetic acid, diesel, biodiesel, and others can be used for autothermal reforming process. The feedstocks, catalysts, and production of gas during steam reforming, dry reforming, and auto thermal reforming are approximately the same despite the difference in the operating condition of these processes. Oxygen in feedstocks can change the transformation route and reduces carbon decomposition by oxidation. In this process, the feed CH_4:O_2:H_2O with a molar ratio of 1:0.33:3 using Pt/Al_2O_3 had shown a maximum conversion of 72% with a H_2 yield of 0.4 (Borgognoni and Tosti,

2012), whereas Ni/Al_2O_3 catalyst had a higher conversion of 96% with H_2 selectivity of 70.4% with a feed ratio of CH_4:Air:H_2O as 1:1:2. These processes are performed at 600–950°C under 1 atm pressure.

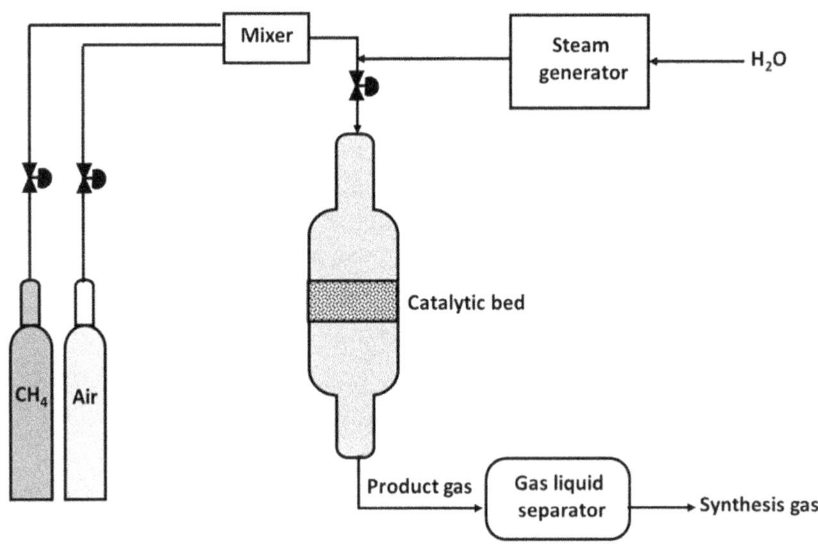

FIGURE 3.3 Schematic of autothermal reforming proccss.

3.6 COMBINED REFORMING OF METHANE

Combined reforming consists of both the steam and dry reforming of methane for the production of syngas (Lachén et al., 2019). Figure 3.4 shows the schematic representation of the combined reforming technology. H_2/CO in the syngas can be maintained by adjusting the molar ratio of $H_2O/CO_2/CH_4$.

$$3CH_4 + CO_2 + 2H_2O \rightleftharpoons 4CO + 8H_2 \tag{3.11}$$

The major advantages of combined steam reforming: (i) minimization of carbon deposition on the catalyst surface, and (ii) consumption of major greenhouse gases (Karemore et al., 2016). One of the major drawbacks of the process is the highly endothermic nature of the reaction and requires very high operating temperatures to achieve the desired conversion. It is reported that a 98% conversion of methane can be achieved using steam and CO_2 at 600°C (Dan et al., 2020).

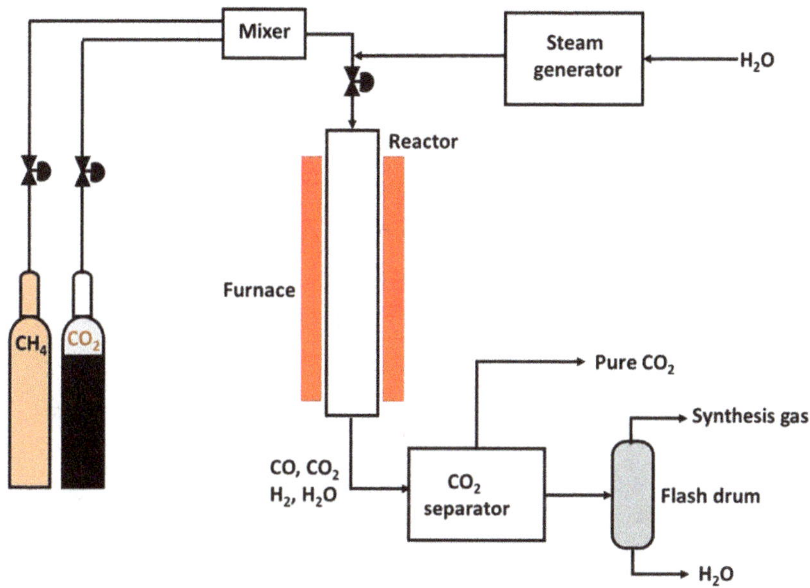

FIGURE 3.4 Schematic diagram of combined steam reforming.

3.7 TRI-REFORMING OF METHANE

Tri-reforming of methane is the combination of steam reforming, dry reforming, and partial oxidation of methane (García-Vargas et al., 2012; Zhang et al., 2013). Figure 3.5 shows the schematic diagram of the tri-reforming of methane. All three processes can be performed in a single reactor for the efficient production of synthesis gas. In tri-reforming of methane, partial oxidation is exothermic. This reaction takes place in a catalytic bed and the heat generated during this reaction is used by the two highly endothermic reactions. Metallic supports are used in the reactors for better transfer of heat to control the temperature within the catalytic bed (Sheng et al., 2011).

Table 3.4 shows the summary of the conversion and selectivity of the tri-reforming of methane. In this process, Ni with Mg, CeO_2 and aerogel can be used with the operating temperatures in the range of 700–800°C. Hydrogen selectivity of 70–78% is reported for Ni-alumina aerogel catalysts.

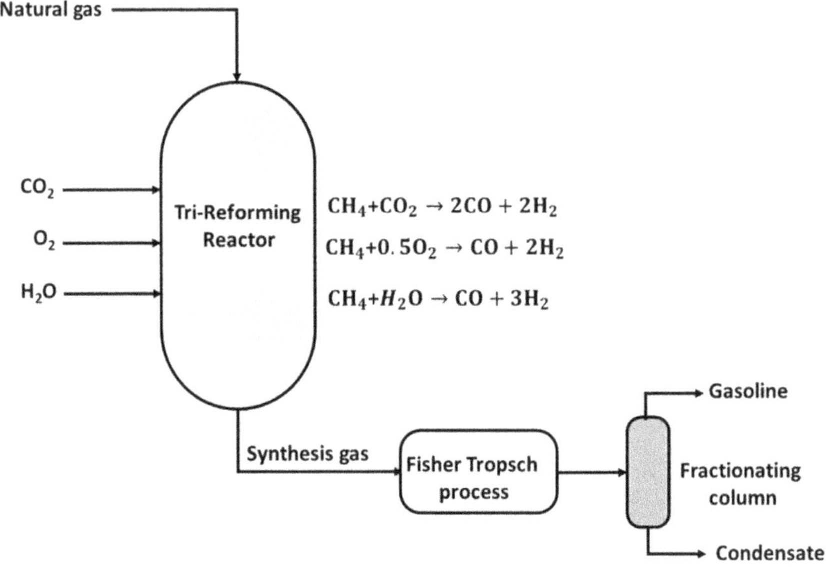

FIGURE 3.5 Schematic diagram of tri-reforming of methane.

3.8 CHEMICAL LOOPING REFORMING

Chemical looping reforming is a novel advanced process to produce hydrogen-rich syngas from different feedstocks through gasification, pyrolysis, methane, and alcohol (Isarapakdeetham et al., 2020). Figure 3.6 shows the schematic representation of the chemical looping reforming technology. In this process, metal oxides are circulated in between three reactors to avoid the cost of air separation. In the first step of the process, fuel, and oxygen-carrying material are fed to a fuel reactor and they react according to the following reaction.

$$(n + m/2)M_xO_y + C_nH_{2m} \rightleftharpoons (n + m/2)M_xO_{y-2} + mH_2O + nCO_2 \qquad (3.12)$$

Where M_xO_y is metal oxides and M_xO_{y-2} is the reduced form of metal oxides. After the reduction of metal oxides, the material is transport to the steam reformer, where steam is used to oxidize the reduced metal oxides with the production of pure hydrogen.

$$(n + m/2)M_xO_{y-2} + mH_2O \rightleftharpoons (n + m/2)M_xO_{y-1} + mH_2 \qquad (3.13)$$

TABLE 3.4 Tri-Reforming of Methane for Hydrogen Production

Precursor (ratio)	Catalyst	Temperature (°C)	Conversion (%)	H_2 selectivity (%)	Reactors	Products	References
CH_4:CO_2:H_2O:O_2:N_2 (24:18:18:3.5:36.5)	Ni/MgO	750	CH_4: 75 CO_2: 80		Fixed bed	CO, H_2	Fedorova et al. (2019)
CH_4:CO_2:H_2O:O_2 (1:0.56:0.46:0.1)	Ni/CeO_2	800	CH_4: 93 CO_2: 83		Fixed bed	CO, H_2	Pino et al. (2011)
	Ni-La/ CeO_2	800	CH_4: 96 CO_2: 86		Fixed bed	CO, H_2	
CH_4:CO_2:H_2O: O_2	NAX (nickel alumina aerogel catalyst)	700	CH_4: 77	70	Fixed bed	CO, H_2	Yoo et al. (2015)
	NAA (nickel alumina aerogel catalyst)	700	CH_4: 83	79	Fixed bed	CO, H_2	

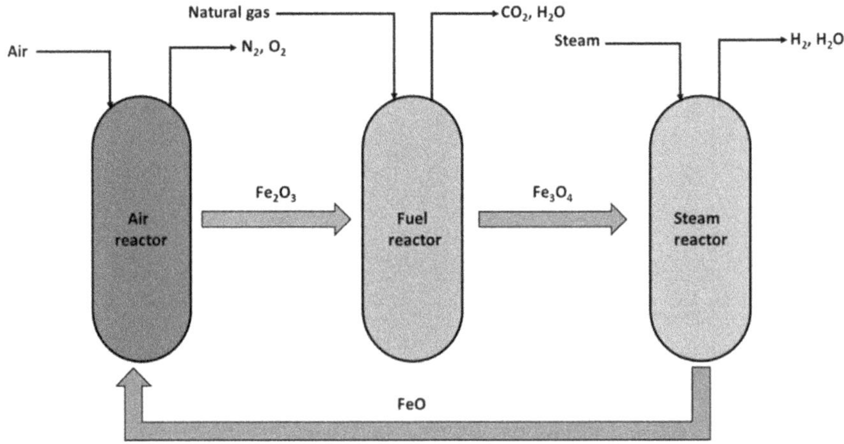

FIGURE 3.6 Schematic diagram of the chemical looping reforming process.

Many researchers show that steam alone is not sufficient for the re-oxidation of metal oxides in the steam reactor so that an air reactor should be added for complete oxidation of the metal oxides as per the following reaction (Shen et al., 2019).

$$(n + m/2)M_xO_{y-1} + 0.5\ O_2 \rightleftharpoons (n + m/2)M_xO_y \quad (3.14)$$

TABLE 3.5 Hydrogen Production by Chemical Looping Reforming Process

Precursors	Catalyst (ratio)	Temperature (°C)	Conversion (%)	Product H_2:CO ratio	Reference
CH_4, steam, air	CuO-Fe_2O_3-Al_2O_3 (20:60:20)	900	88.5	1.5	Nadgouda et al. (2019)
	CuO-Fe_2O_3-Al_2O_3 (40:40:20)	900	76.3	1.6	
	CuO-Fe_2O_3-Al_2O_3 (60:20:20)	900	69	1.45	

Hydrogen obtained from the chemical looping reforming process depends on various conditions such as the type of feed used, reactor type, oxygen-carrying material and operating condition of the reactor (Spragg et al., 2018). One of the major challenges in the design of the chemical-looping reforming system is to integrate three reactors for the circulation of metal particles. In addition, heat balancing between the reactors is difficult due to the highly exothermic and endothermic nature of reactions in air reactor and fuel reactors, respectively (Jiang et al., 2020). Table 3.5 summarizes a specific study on chemical looping reforming of hydrogen production at 900°C using CuO-Fe_2O_3-Al_2O_3 based catalyst. It is reported that 88.5% conversion of methane can be achieved with the production of ultrapure hydrogen.

3.9 PYROLYSIS

Pyrolysis is a treatment in which solid fuels are treated in the absence of oxygen. Figure 3.7 shows the schematic diagram of a pyrolysis process. This process requires the operating temperature between 600°C and 1000°C to produce various products in the form of gas, liquid, and char. The gas products mainly contain CO_2, CO, and H_2. The liquid contents are pitch tar, light oil and heavy oil. The composition of gaseous components mainly depends upon the nature of biomass, reactor type, its operating parameter, and the residence time of the feed in the reactor (Soria et al., 2019)

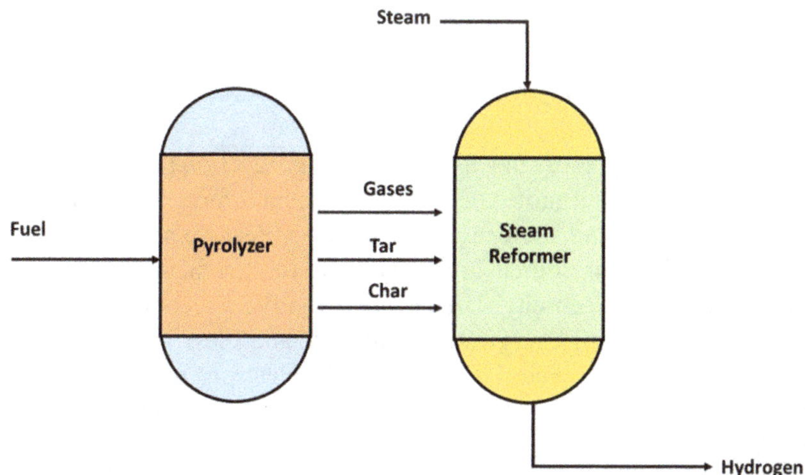

FIGURE 3.7 Schematic diagram of the two-stage pyrolysis-steam reforming process.

The generated product gas and bio-oil are energy-intensive compared to the raw biomass. These bio-oil and gases can be used as a feed for the production of acetic acid, methanol, and synthesis gas (Czernik and Bridgwater, 2004; Akubo et al., 2019). In addition, the pyrolysis process followed by the catalytic steam reforming process directly converts the product gas into hydrogen. Table 3.6 shows the catalysts and conversion achieved during the pyrolysis reaction. It can be seen that a hydrogen selectivity in the range of 44 to 49% can be obtained at the operating temperatures between 500°C and 800°C. It can be noted that Cr_2O_3 yielded a slightly higher selectivity of 49% compared to other catalysts.

3.10 GASIFICATION

Gasification is a thermochemical partial oxidation process that produces CO_2, CO, H_2, and other light and heavier hydrocarbons. This process is carried out at high temperatures with a controlled amount of oxygen or air (Nanda et al., 2014). The process uses a catalyst for hydrogenation in the temperature range of 800–1200°C. The generated synthesis gas or producer gas in this process can be used to produce electricity and chemicals (Sikarwar et al., 2016). This is a promising technology as it reduces the mass of biomass by 70% and volume by 90%. Table 3.7 shows the selectivity of hydrogen production through biomass gasification. Nearly 57–64% of hydrogen selectivity is estimated with nickel as the catalyst at the operating temperatures between 600°C and 950°C.

3.11 WATER SPLITTING METHOD

Water splitting is a process in which hydrogen production takes place by the direct splitting of water molecules into its elemental form. In water, H-O-H bonds are present, and the energy required to break this bond is mainly supplied by different sources such as light (electromagnetic radiation), thermal (heat) and electricity. The difference in the water-splitting process is based on the type of energy source applied, and these are referred to as photolysis, thermolysis, and electrolysis. In the electrolysis process, oxygen, and proton form at the anode and hydrogen form at the cathode (Varadhan et al., 2017).

CO_2 generation is zero, and thus this process can be eco-friendly. However, the electricity used in this process is generated by the combustion of coal or

TABLE 3.6 Hydrogen Selectivity Under Catalytic Pyrolysis of Biomass Fuels

Precursors	Catalyst	Temperature (°C)	H_2 Selectivity	Reactor	Products	References
Sawdust	• NiO • CuO • Al_2O_3 • Cr_2O_3	500–800	• 45 • 45 • 45.6 • 48	Fixed bed	H_2, CO, CO_2, H_2O, CH_4	Chen et al. (2003)
Rice straw	• NiO • Al_2O_3 • MnO • Cr_2O_3	500–800	• 44.2 • 45 • 48.5 • 49.3	Fixed bed	H_2, CO, CO_2, H_2O, CH_4, Char	

TABLE 3.7 Hydrogen Selectivity and Operating Conditions of Catalytic Biomass Gasification

Precursors	Catalyst	Temp. (°C)	H_2 selectivity (%)	Reactors	Steam-to-carbon ratio	Products	References
Wheat straw	Ni	600–900	64	Fixed bed	0–3	CO, H_2 and others	Yao et al. (2016)
Sawdust	Ni	750–950	57.4	Fluidized bed	0–3	CO, H_2 and others	Turn (1998)

TABLE 3.8 Hydrogen Production by Water Splitting Reaction

Precursors	Catalyst	Electric power supplied (Watt)	Power source	Average evolution rate (µmol)/g	Reaction time (h)	References
Water	• TiO_2	• 300	Xe-lamp	• 257	• 4	
	• N-TiO_2	• 300		• 600	• 4	Naik et al. (2019)
	• Au/TiO_2	• 300		• 710	• 4	
	• Au-N/TiO_2	• 300		• 976	• 4	
Water	• Cu/ TiO_2	• 300	Xe-lamp	• 740	• 4	Liu et al. (2018)
Water	• TiO_2	• 300	Xe-lamp	• 140	• 4	Wang et al. (2017)
	• 2Fe/TiO_2	• 300	Xe-lamp	• 697	• 4	

natural gas which produces CO_2 (Fajrina and Tahir, 2019). Photocatalytic water splitting can be a better choice to produce hydrogen by using a renewable source of energy such as sunlight and water. The production of hydrogen by this process helps in reducing the global warming effect by lowering the CO_2 production. In this process, water undergoes a redox reaction using a photo-catalyst, which can be chosen in such a way that it can be a good absorber of sunlight energy to produce a significant quantity of hydrogen. This process requires better interaction between catalyst, reactant, and light (Chouhan et al., 2016). The primary requirement of the photocatalytic process is light energy, which must be higher or equal than the bond energy gap of the semiconductor-based photo-catalyst (Acar et al., 2014). The thermolysis process is another method to produce hydrogen by water splitting using heat as an energy source. This process requires high temperatures (500–1000°C) to produce hydrogen (Muhich et al., 2015). Table 3.8 shows the comparison of hydrogen production through the water-splitting reaction. It can be noted that Au-N/TiO_2 catalyst yielded the highest hydrogen evolution rate using a 300-watt power Xe-lamp.

3.12 CONCLUSIONS

The steam reforming process using various kinds of feedstock and catalyst produces hydrogen. Methane, ethane, acetic acid, methanol, ethanol, methane, ethane, tar are the potential feedstocks for hydrogen production through the catalytic reforming process. Ethanol exhibited a good reforming tendency with steam for hydrogen generation using nickel-based catalysts. Dry reforming using CO_2 and methane could be a viable option. However, it requires high operating temperatures. The partial oxidation method can be a better option as this process generates its required energy, but the formation of hotspots deteriorates the activity of catalysts. Tri-reforming and chemical looping reforming are the emerging technologies to produce hydrogen. Chemical looping reforming technology can produce ultrapure hydrogen directly from the steam without any contamination from other gasses. However, the integration of three reactors and the circulation of metal oxides among them and heat balance between the reactors are the major challenges for hydrogen production. Other alternate technologies such as pyrolysis, gasification, and water splitting, etc., can be used to produce hydrogen. In all the processes, noble metals or transition metals are the catalysts; noble metals are highly efficient catalysts as it resists the deposition of carbon;

however, it is expensive. Alternatively, transition metals are low cost but it gets deactivate with carbon deposition. Hence, appropriate catalyst development is important for the efficient conversion of feedstocks to hydrogen.

KEYWORDS

- **autothermal reforming**
- **catalytic conversion**
- **chemical looping reforming**
- **combined reforming of methane**
- **dry reforming**
- **dry reforming of methane**
- **hydrogen**
- **partial oxidation**
- **steam reforming**
- **tri-reforming of methane**

REFERENCES

Abdullah, N., Ainirazali, N., Chong, C. C., Razak, H. A., Setiabudi, H. D., Jalil, A. A., & Vo, D. V. N., (2020). Influence of impregnation assisted methods of Ni/SBA-15 for production of hydrogen via dry reforming of methane. *Int. J. Hydrogen Energy, 45*, 18426–18439.

Abdulrasheed, A. A., Jalil, A. A., Hamid, M. Y. S., Siang, T. J., Fatah, N. A. A., Izan, S. M., & Hassan, N. S., (2020). Dry reforming of methane to hydrogen-rich syngas over robust fibrous KCC-1stabilized nickel catalyst with high activity and coke resistance. *Int. J. Hydrogen Energy, 45*, 18549–18561.

Acar, C., Dincer, I., & Zamfirescu, C., (2014). A review on selected heterogeneous photocatalysts for hydrogen production. *Int. J. Energy Res., 38*, 1903–1920.

Akubo, K., Nahil, M. A., & Williams, P. T., (2019). Pyrolysis-catalytic steam reforming of agricultural biomass wastes and biomass components for production of hydrogen/syngas. *J. Energy. Inst., 92*, 1987–1996.

Amin, A. M., Croiset, E., & Epling, W., (2011). Review of methane catalytic cracking for hydrogen production. *Int. J. Hydrogen Energy, 36*, 2904–2935.

Antaleo, G., Parola, V. L., Deganello, F., Singha, R. K., Bal, R., & Venezia, A. M., (2016). Ni/CeO_2 catalysts for methane partial oxidation: Synthesis driven structural and catalytic effects. *Appl. Catal. B. Environ., 189*, 233–241.

Ball, M., & Weeda, M., (2015). The hydrogen economy - vision or reality. *Int. J. Hydrogen Energy, 40*, 7903–7919.

Bang, Y., Han, S. J., Yoo, J., Choi, J. H., Kang, K. H., Song, J. H., & Song, I. K., (2013). Hydrogen production by steam reforming of liquefied natural gas (LNG) over trimethyl benzene-assisted ordered mesoporous nickel-alumina catalyst. *Int. J. Hydrogen Energy, 38*, 8751–8758.

Bang, Y., Seo, J. G., Youn, M. H., & Song, I. K., (2012). Hydrogen production by steam reforming of liquefied natural gas (LNG) over mesoporous Ni-Al_2O_3 aerogel catalyst prepared by a single-step epoxide-driven sol-gel method. *Int. J. Hydrogen Energy, 37*, 1436–1443.

Basagiannis, A. C., & Verykios, X. E., (2007). Catalytic steam reforming of acetic acid for hydrogen production. *Int. J. Hydrogen Energy, 32*, 3343–3355.

Bizkarra, K., Barrio, V. L., Gartzia-Rivero, L., Banuelos, J., Lopez-Arbeloa, I., & Cambra, F., (2019). Hydrogen production from a model bio-oil/bioglycerol mixture through steam reforming using Zeolite L supported catalysts. *Int. J. Hydrogen Energy, 44*, 1492–1504.

Borgognoni, F., & Tosti, S., (2012). Pd-Ag multi-membranes module for hydrogen production by methane auto-thermal reforming. *Int. J. Hydrogen Energy, 37*, 1444–1453.

Cai, F., Lu, P., Ibrahim, J., Fu, Y., Zhang, J., & Sun, Y., (2018). Investigation of the role of Nb on Pd-Zr-Zn catalyst in methanol steam reforming for hydrogen production. *Int. J. Hydrogen Energy, 4*, 11717–11733.

Chen, D., Wang, W., & Liu, C., (2020). Hydrogen production through glycerol steam reforming over beehive-biomimetic graphene-encapsulated nickel catalysts. *Renew. Energy, 145*, 2647–2657.

Chen, G., Andries, J., & Spliethoff, H., (2003). Catalytic pyrolysis of biomass for hydrogen rich fuel gas production. *Energy Convers. Manag., 44*, 2289–2296.

Chouhan, N., Ameta, R., Meena, R. K., Mandawat, N., & Ghildiyal, R., (2016). Visible light-harvesting Pt/CdS/Co-doped ZnO nanorods molecular device for hydrogen generation. *Int. J. Hydrogen Energy, 41*, 2298–2306.

Corbo, P., & Migliardini, F., (2007). Hydrogen production by catalytic partial oxidation of methane and propane on Ni and Pt catalysts. *Int. J. Hydrogen Energy, 32*, 55–66.

Czernik, S., & Bridgwater, A. V., (2004). Overview of applications of biomass fast pyrolysis oil. Energy *Fuel, 18*, 590–598.

Da Costa-Serra, J. F., & Chica, A., (2018). Catalysts based on co-birnessite and co-todorokite for the efficient production of hydrogen by ethanol steam reforming. *Int. J. Hydrogen Energy, 43*, 16859–16865.

Dan, M., Mihet, M., & Lazar, M. D., (2020). Hydrogen and/or syngas production by combined steam and dry reforming of methane on nickel catalysts. *Int. J. Hydrogen Energy, 45*, 26254–26264.

Dodds, P. E., & McDowall, W., (2013). The future of the UK gas network. *Energ. Policy, 60*, 305–316.

Evans, S. E., Staniforth, J. Z., Darton, R. J., & Ormerod, R. M., (2014). A nickel doped perovskite catalyst for reforming methane rich biogas with minimal carbon deposition. *Green Chem., 16*, 4587–4594.

Fajrina, N., & Tahir, M., (2019). A critical review in strategies to improve photocatalytic water splitting towards hydrogen production. *Int. J. Hydrogen Energy, 44*, 540–577.

Fedorova, Z. A., Danilova, M. M., & Zaikovskii, V. I., (2019). Porous nickel-based catalysts for tri- reforming of methane to synthesis gas: Catalytic activity. *Mater., 261*, 127087.

García-Diéguez, M., Pieta, I. S., Herrera, M. C., Larrubia, M. A., & Alemany, L. J., (2010). Nanostructured Pt- and Ni-based catalysts for CO_2-reforming of methane. *J. Catal, 270*, 136–145.

García-Vargas, J. M., Valverde, J. L., De Lucas-Consuegra, A., Gómez-Monedero, B., Sánchez, P., & Dorado, F., (2012). Precursor influence and catalytic behaviour of Ni/CeO_2 and Ni/SiC catalysts for the tri-reforming process. *Appl. Catal. A-Gen., 431, 432*, 49–56.

Hambali, H. U., Jalil, A. A., Abdul, R. A. A., Siang, T. J., & Vo, D.V. N., (2020). Enhanced dry reforming of methane over mesostructured fibrous Ni/MFI zeolite: Influence of preparation methods. *J. Energy. Inst., 93*, 1535–1543.

Houteit, A., Mahzoul, H., Ehrburger, P., Bernhardt, P., Légaré, P., & Garin, F., (2006). Production of hydrogen by steam reforming of methanol over copper-based catalysts: The effect of cesium doping. *Appl. Catal A: Gen., 306*, 22–28.

Hu, X., & Lu, G., (2010). Comparative study of alumina-supported transition metal catalysts for hydrogen generation by steam reforming of acetic acid. *Appl. Catal. B.: Environ., 99*, 289–297.

Hwang, B. Y., Sakthinathan, S., & Chiu, T. W., (2019). Production of hydrogen from steam reforming of methanol carried out by self-combusted $CuCr_{1-x}Fe_xO_2$ (x = 0–1) nano-powders catalyst. *Int. J. Hydrogen Energy, 44*, 2848–2856.

Isarapakdeetham, S., Kim-Lohsoontorn, P., Wongsakulphasatch, S., Kiatkittipong, W., Laosiripojana, N., Gong, J., & Assabumrungrat, S., (2020). Hydrogen production via chemical looping steam reforming of ethanol by Ni-based oxygen carriers supported on CeO_2 and La_2O_3 promoted Al_2O_3. *Int. J. Hydrogen Energy, 45*, 1477–1491.

Jiang, B., Li, L., Qian, Z., Ma, J., Zhang, H., Bai, J., & Tang, D., (2020). Chemical looping reforming of glycerol for continuous H_2 production by moving-bed reactors: Simulation and experiment. *Energy Fuel, 34*, 1841–1850.

Karemore, A. L., Vaidya, P. D., Sinha, R., & Chugh, P., (2016). On the dry and mixed reforming of methane over Ni/Al_2O_3 - Influence of reaction variables on syngas production. *Int. J. Hydrogen Energy, 41*, 22963–22975.

Kumar, A., Singh, R., & Sinha, A. S. K., (2019). Catalyst modification strategies to enhance the catalyst activity and stability during steam reforming of acetic acid for hydrogen production. *Int. J. Hydrogen Energy, 44*, 12983–1301.

Lachén, J., Herguido, J., & Peña, J. A., (2019). Production and purification of hydrogen by biogas combined reforming and steam-iron process. *Int. J. Hydrogen Energy, 44*, 19244–19254.

Li, L., Zuo, S., An, P., Wu, H., Hou, F., Li, G., & Liu, G., (2020). Hydrogen production via steam reforming of n-dodecane over Ni Pt alloy catalysts. *Fuel, 262*, 116469.

Liu, Q. F., Zhang, Q., Liu, B. R., Li, S., & Ma, J. J., (2018). Building surface defects by doping with transition metal on ultrafine TiO_2 to enhance the photocatalytic H_2 production activity. *Chinese J. Catal., 39*, 542–548.

Ma, Y., Ma, Y., Zhao, Z., Hu, X., Ye, Z., Yao, J., & Dong, D., (2019). Comparison of fibrous catalysts and monolithic catalysts for catalytic methane partial oxidation. *Renew. Energy, 138*, 1010–1017.

Marbán, G., & Valdés-Solís, T., (2007). Towards the hydrogen economy. *Int. J. Hydrogen Energy, 32*, 1625–1637.

Matsumura, Y., & Nakamori, T., (2004). Steam reforming of methane over nickel catalysts at low reaction temperature. *Appl. Catal. A: Gen., 258*, 107–114.

Muhich, C. L., Ehrhart, B. D., Al-Shankiti, I., Ward, B. J., Musgrave, C. B., & Weimer, A. W., (2015). A review and perspective of efficient hydrogen generation via solar thermal water splitting. *Wiley Interdiscip. Rev.: Energy Environ., 5*, 261–287.

Nadgouda, S. G., Guo, M., Tong, A., & Fan, L. S., (2019). High purity syngas and hydrogen co-production using copper-iron oxygen carriers in chemical looping reforming process. *Applied Energy, 235*, 1415–1426.

Naik, G. K., Majhi, S. M., Jeong, K. U., Lee, I. H., & Yu, Y. T., (2019). Nitrogen doping on the core-shell structured Au@TiO_2 nanoparticles and its enhanced photocatalytic hydrogen evolution under visible light irradiation. *J. Alloys Compd., 771*, 505–512.

Nanda, S., Mohammad, J., Reddy, S. N., Kozinski, J. A., & Dalai, A. K., (2014). Pathways of lignocellulosic biomass conversion to renewable fuels. *Biomass Conv. Bioref., 4*, 157–191.

Nanda, S., Rana, R., Zheng, Y., Kozinski, J. A., & Dalai, A. K., (2017). Insights on pathways for hydrogen generation from ethanol. *Sustain. Energy Fuels, 1*, 1232–1245.

Nguyen, D. D., Ngo, S. I., Lim, Y. I., Kim, W., Lee, U. D., Seo, D., & Yoon, W. L., (2019). Optimal design of a sleeve-type steam methane reforming reactor for hydrogen production from natural gas. *Int. J. Hydrogen Energy, 44*, 1973–1987.

Noh, Y. S., Lee, K. Y., & Moon, D. J., (2019). Hydrogen production by steam reforming of methane over nickel based structured catalysts supported on calcium aluminate modified SiC. *Int. J. Hydrogen Energy, 44*, 21010–21019.

Osman, A. I., (2020). Catalytic hydrogen production from methane partial oxidation: Mechanism and kinetic study. *Chem. Eng. Technol., 43*, 641–648.

Osman, A. I., Meudal, J., Laffir, F., Thompson, J., & Rooney, D., (2017). Enhanced catalytic activity of Ni on η-Al_2O_3 and ZSM-5 on addition of ceria-zirconia for the partial oxidation of methane. *Appl. Catal. B. Environ., 212*, 68–79.

Özkan, G., Şahbudak, B., & Özkan, G., (2018). Effect of molar ratio of water / ethanol on hydrogen selectivity in catalytic production of hydrogen using steam reforming of ethanol. *Int. J. Hydrogen Energy, 44*, 9823–9829.

Özkara-Aydınoğlu, Ş., & Aksoylu, A. E., (2010). Carbon dioxide reforming of methane over Co-X/ZrO_2 catalysts (X=La, Ce, Mn, Mg, K). *Catal., 11*, 1165–1170.

Pino, L., Vita, A., Cipitì, F., Laganà, M., & Recupero, V., (2011). Hydrogen production by methane tri-reforming process over Ni-ceria catalysts: Effect of La-doping. *Appl. Catal. B Environ., 104*, 64–73.

Rodrigues, T. S., De Moura, A. B. L., Silva, F. A., Candido, E. G., Da Silva, A. G. M., De Oliveira, D. C., & Fonseca, F. C., (2019). Ni supported $Ce_{0.9}Sm_{0.1}O_{2-\delta}$ nanowires: An efficient catalyst for ethanol steam reforming for hydrogen production. *Fuel, 237*, 1244–1253.

Sarangi, P. K., & Nanda, S., (2020). Biohydrogen production through dark fermentation. *Chem. Eng. Technol., 43*, 601–612.

Seo, J. G., Youn, M. H., Bang, Y., & Song, I. K., (2011). Hydrogen production by steam reforming of simulated liquefied natural gas (LNG) over mesoporous nickel-M-alumina (M=Ni, Ce, La, Y, Cs, Fe, Co, and Mg) aerogel catalysts. *Int. J. Hydrogen Energy, 36*, 3505–3514.

Shanmugam, V., Neuberg, S., Zapf, R., Pennemann, H., & Kolb, G., (2020). Hydrogen production over highly active Pt based catalyst coatings by steam reforming of methanol: Effect of support and co-support. *Int. J. Hydrogen Energy, 45*, 1658–1670.

Shen, Y., Zhao, K., He, F., & Li, H., (2019). The structure-reactivity relationships of using three-dimensionally ordered macroporous $LaFe_{1-x}Ni_xO_3$ perovskites for chemical-looping steam methane reforming. *J. Energy Inst., 92*, 239–246.

Sheng, M., Yang, H., Cahela, D. R., & Tatarchuk, B. J., (2011). Novel catalyst structures with enhanced heat transfer characteristics. *J. Catal., 281*, 254–262.

Siang, T. J., Singh, S., Omoregbe, O., Bach, L. G., Phuc, N. H. H., & Vo, D. V. N., (2018). Hydrogen production from CH_4 dry reforming over bimetallic Ni-Co/Al_2O_3 catalyst. *J. Energy Inst., 91*, 683–694.

Sikarwar, V. S., Zhao, M., Clough, P., Yao, J., Zhong, X., Memon, M. Z., & Fennell, P. S., (2016). A overview of advances in biomass gasification. *Energy Environ. Sci., 9*, 2939–2977.

Soria, M. A., Barros, D., & Madeira, L. M., (2019). Hydrogen production through steam reforming of bio-oils derived from biomass pyrolysis: Thermodynamic analysis including in situ CO_2 and/or H_2 separation. *Fuel, 244*, 184–195.

Spragg, J., Mahmud, T., & Dupont, V., (2018). Hydrogen production from bio-oil: A thermodynamic analysis of sorption-enhanced chemical looping steam reforming. *Int. J. Hydrogen Energy, 43*, 22032–22045.

Sun, Y., Zhang, G., Liu, J., Xu, Y., & Lv, Y., (2020). Production of syngas via CO_2 methane reforming process: Effect of cerium and calcium promoters on the performance of Ni-MSC catalysts. *Int. J. Hydrogen Energy, 45*, 640–649.

Tana, R. S., Abdullaha, T. A. T., Ripina, A., Ahmada, A., & Md Isa, K., (2019). Hydrogen-rich gas production by steam reforming of gasified biomass tar over Ni/dolomite/La_2O_3 catalyst. *J. Environ. Chem. Eng., 7*, 103490.

Tuna, C. E., Silveira, J. L., Da Silva, M. E., Boloy, R. M., Braga, L. B., & Pérez, N. P., (2018). Biogas steam reformer for hydrogen production: Evaluation of the reformer prototype and catalysts. *Int. J. Hydrogen Energy, 43*, 2108–2120.

Turn, S., (1998). An experimental investigation of hydrogen production from biomass gasification. *Int. J. Hydrogen Energy, 23*, 641–648.

Varadhan, P., Fu, H. C., Priante, D., Retamal, J. R. D., Zhao, C., Ebaid, M., & He, J. H., (2017). Surface passivation of GaN nanowires for enhanced photo-electrochemical water-splitting. *Nano Lett., 17*, 1520–1528.

Vozniuk, O., Tanchoux, N., Millet, J. M., Albonetti, S., Di Renzo, F., & Cavani, F., (2019). Spinel mixed oxides for chemical-loop reforming: From solid state to potential application. horizons in sustainable industrial chemistry and catalysis. *Stud. Surf. Sci. Catal.*, 281–302.

Wang, F., Shen, T., Fu, Z., Lu, Y., & Chen, C., (2017). Enhanced photocatalytic water-splitting performance using Fe-doped hierarchical TiO_2 ball-flowers. *Nanotechnology, 29*, 035702.

Wang, F., Zhang, L., Deng, J., Zhang, J., Han, B., Wang, Y., & Deng, Z., (2019). Embedded Ni catalysts in Ni-O-Ce solid solution for stable hydrogen production from ethanol steam reforming reaction. *Fuel Process. Technol., 193*, 94–101.

Yadav, A. K., & Vaidya, P. D., (2019). Renewable hydrogen production by steam reforming of butanol over multi-walled carbon nanotube-supported catalysts. *Int. J. Hydrogen Energy, 44*, 30014–30023.

Yao, D., Hu, Q., Wang, D., Yang, H., Wu, C., Wang, X., & Chen, H., (2016). Hydrogen production from biomass gasification using biochar as a catalyst/support. *Bioresour. Technol., 216*, 159–164.

Yoo, J., Bang, Y., Han, S. J., Park, S., Song, J. H., & Song, I. K., (2015). Hydrogen production by trireforming of methane over nickel-alumina aerogel catalyst. *J. Mol. Catal. A Chem., 410*, 74–80.

Yoo, J., Park, S., Song, J. H., Yoo, S., & Song, I. K., (2017). Hydrogen production by steam reforming of natural gas over butyric acid-assisted nickel/alumina catalyst. *Int. J. Hydrogen Energy, 42*, 28377–28385.

Zhang, C., Hu, X., Yu, Z., Zhang, Z., Chen, G., Li, C., & Hu, S., (2019). Steam reforming of acetic acid for hydrogen production over attapulgite and alumina supported Ni catalysts: Impacts of properties of supports on catalytic behaviors. *Int. J. Hydrogen Energy, 44*, 5230–5244.

Zhang, Q., Zhang, T., Shi, Y., Zhao, B., Wang, M., Liu, Q., & Ning, P., (2017). A sintering and carbon resistant Ni-SBA-15 catalyst prepared by solid-state grinding method for dry reforming of methane. *J.CO_2 Util., 17*, 10–19.

Zhang, Y., Cruz, J., Zhang, S., Lou, H. H., & Benson, T. J., (2013). Process simulation and optimization of methanol production coupled to tri-reforming process. *Int. J. Hydrogen Energy, 38*, 13617–13630.

CHAPTER 4

Co-Conversion of Plastic Wastes and Biomass into Biohydrogen

KRUSHNA PRASAD SHADANGI

Department of Chemical Engineering, Veer Surendra Sai University of Technology, Sambalpur, Odisha, India
E-mail: kpshadangi_chemical@vssut.ac.in

ABSTRACT

Plastic wastes are usually not biodegradable. Nowadays, plastics are used everywhere, which increases their concentration in municipal solid waste. Since the landfill of waste plastics is not environment friendly, their utilization for the production of energy is the best option. Biomass is also abundantly available. The source of biomass and waste plastics are the best source to produce alternative fuel. Gasification is one of the thermochemical techniques that produce several combustible gasses. Among all the combustible gasses, hydrogen is a fuel that does not pollute the environment when burnt. However, individual gasification of different plastics creates various problems related to the melting and formation of char. The yield of H_2 is also not so attractive due to the formation of CO, CO_2, CH_4, and various low carbon hydrocarbon fuels as well. Hence, the co-gasification of waste plastics with biomass can be a good choice where the waste biomass can be used for the production of H_2 energy. The efficiency of the production of H_2 gas depends on the biomass to plastic ratio, gasification agents, types of catalysts and temperature. Hence, optimization of the process to produce a higher yield of H_2 is precise during the co-gasification process.

4.1 INTRODUCTION

Hydrogen (H_2) is one of the sources of future energy that provides better heating value compared to fossil fuels. The calorific value of H_2 (120 MJ/kg) is more than two times of non-conventional fuels (45 MJ/kg) such as diesel and petrol. Hydrogen has the utmost energy density compared to other fuels (Parthasarathy et al., 2014). The combustion of H_2 gas produces heat energy and water, whereas fossil fuel emits various greenhouse gasses such as SO_x, NO_x, and CO_2 upon combustion. Hence, H_2 is known as a green fuel. Hydrogen fuel is called green energy since H_2 combustion does not produce any harmful emissions. Hydrogen can be converted into ethanol and methanol and be used directly in the internal combustion engine and fuel cell (Sarangi and Nanda, 2020; Nanda et al., 2017). H_2 is also used in the petroleum industry as a petrochemical feedstock and as a reduction agent for steel plants. The storage and transportation of H_2 are also easy. Liquid hydrocarbons, natural gas, and coal are generally used to produce H_2 commercially.

Plastic wastes and biomass are such feedstocks that can be converted into hydrogen and other biofuels through several thermochemical technologies (Nanda and Berruti, 2021a,b). Since large quantities of plastic wastes and waste biomass are generated, their utilization for the production of green energy is a promising option for waste management, which will not affect the habitation. Various conversion techniques are being followed to convert the biomass and plastic waste to H_2 energy such as pyrolysis and gasification. These thermochemical conversion techniques can convert the plastic waste to H_2.

According to the literature, the yield of H_2 is higher in the gasification process compared to pyrolysis (Chunfei and Paul, 2010). It was also reported that the two-stage gasification system has more impact on the production of higher yield of H_2 compared to the single-stage gasification process. However, catalytic pyrolysis and gasification resulted in better than thermal pyrolysis concerning the yield of H_2. The study also reported that the yield of H_2 varied with temperature, the types of pyrolysis, catalyst, and catalytic composition.

Kumagai et al. (2015) revealed the effect of pyrolysis-gasification of biomass/plastic mixture on the yield of H_2, such as the yield varied with temperature, catalytic ratio and temperature. This study confirmed that not only the temperature and composition of the catalyst but also the feed stick composition affects the yield of H_2. Hence, the co-conversion of plastic waste and biomass into hydrogen is a very interesting topic. In this chapter, the co-conversion of plastic waste and biomass into hydrogen along with the effect of various parameters including plastic and biomass ratio, catalytic conversion methods are reflected.

4.2 FEEDSTOCKS FOR HYDROGEN PRODUCTION

Biomass and waste plastics are best for feedstocks for the generation of H_2. Since, both of the sources are a composition of hydrocarbons, although biomass has oxygen, breaking of the carbon-hydrogen bonds can produce H_2 as one of the products. Waste biomass and waste plastics are abundantly available on the earth's surface, which can be collected easily and can cut the cost of the raw materials. Varieties of biomass are available in the world, and some of them are edible and non-edible. The best is the utilization of a non-edible source of biomass for the production of H_2 energy.

Lignocellulosic biomass (e.g., agricultural, and woody biomass) is a composition of cellulose, hemicellulose, lignin, and extractives (Okolie et al., 2021). The biomass containing less extractive is better than others since the extractive can be extracted and can be used to produce bio-oil, which has wide uses such as the production of soap, liquid fuel and can have health benefits also (Shadangi and Mohanty, 2013). Hence, other than non-edible seeds the other parts of a plant along with forestry residues, waste from the food industry, organic municipal waste, agricultural residues and waste from industry can be used for the production of H_2 energy (Kumar et al., 2009; Puig-Arnavat and Coronas, 2010; Ahrenfeldt et al., 2013).

Plastics are the polymer, which does not degrade easily. Hence, the best use is that either it can be recycled or can be used for energy production. In general, all types of plastics such as polyethylene (PE), polypropylene (PP), low-density polyethylene (LDPE), high-density polyethylene (HDPE), polyurethane, polyvinyl chloride (PVC), polyethylene terephthalate (PET) and polystyrene (PS) can be used as feedstocks for H_2 and other biofuel production (Alvarez et al., 2014; Block et al., 2019; Nanda and Berruti, 2021b). Biomass or plastic or a mixture of both can be used as the feedstock for the production of H_2. The preparation of the feedstock for H_2 production is very important since it affects the yield. The dried feedstock, including the biomass and plastics should be homogeneous to enhance the yield of H_2 (Alvarez et al., 2014).

4.3 GASIFICATION

Gasification is a technique that converts all carbon-based materials such as biomass, coal, and plastics into synthesis gas such as CO and H_2. The gasification reaction occurs in a reduced oxygen environment. The process is also known as a partial oxidation process. In the process of gasification, the carbon reacts with steam and oxygen at certain high pressure and temperature and produces gaseous fuel along with few byproducts. These byproducts are removed from the gasifier and used to produce a fuel that can be used in petrochemical industries, production of H_2, and for the generation of electricity or steam. Gasification is the process that converts all types of organic wastes into valuable products (Bridgwater, 2003; Alvarez et al., 2014; Block et al., 2019).

Gasifiers of different designs such as updraft or counter-current gasifier, downdraft or co-current gasifier, cross-draft gasifier, fluidized bed gasifier, double-fired, and entrained bed gasifier are used for the gasification of biomass and plastics (Block et al., 2019). Each type of gasifier has several advantages and disadvantages concerning the fuel type, ease of operation, efficiency, stability, gas quality and applications (Pinto et al., 2002; Robinson et al., 2016). The gasifier types are separated as per their feed inlet, air inlet and gas outlet. The selection of a gasifier is less important. The important parameters that depend are stability, gas quality, gasifier efficiency and pressure losses along with the size of the feed, moisture content, volatile matter content, ash content, energy content, bulk density and chemical composition of ash of the feed sample (Kumar et al., 2009; Robinson et al., 2016). Hence, for individual fuel, it is essential to confirm that the fuel meets the requirements of the gasifier before selecting the gasifier. The moisture in the feed shrinks the thermal efficiency of the gasifier as the heat is used to vaporize the moisture. Thus, sufficient energy is not available for the reduction reactions and for converting the thermal energy into chemical bond energy in the gas.

The higher moisture in feed increases the possibility of formation of more tar at low-temperature combustion, which affects directly the efficiency of the gasifier (Puig-Arnavat and Coronas, 2010). Hence, feed containing less moisture is suitable for gasification. In comparison to updraft gasifier, a downdraft gasifier requires less moisture content feed material. Higher volatile matter in the feed enhances the formation of more gas. Less ash content (5–6%) in feed is more suitable for any type of gasifier since the agglomeration of ashes helps in slagging or clinker formation in the reactor, which leads to the formation of excessive tar in the reactor (Kumar et al.,

2009). Higher the bulk density (weight/unit volume) results in higher gas heating values and helps the burning capacity of the char in the reduction zone. The bulk density can be amplified by pelletizing, which is affected by the particle size of the feed and moisture content as well. The fine sized particles are less suitable for up and downdraft gasifier as it creates flow problems and results in dust in the outlet gas and increase the pressure drop in the reduction zone of the gasifier. Higher the pressure drop enhances the formation of char. However, a relatively higher particle-sized feedstock can be used in a fluidized bed gasifier. If the particle size is excessive large, then there will be a prospect of startup problem, gas channeling problems and yielding poor gas quality. The channeling problem is highly probable in updraft gasifier. The acceptable feed size is influenced by the design of the gasifier (Kumar et al., 2009).

4.4 GASIFICATION REACTION MECHANISM

Gasification is one of the thermochemical conversion processes which converts the organic and carbonaceous materials into CO, H_2, CO_2, CH_4 and C_{2+} gasses in the presence of a controlled amount of steam and oxygen at higher temperatures (>700°C). Gasification is not a single-stage process. The gasification chemical reactions are followed by dehydration (drying zone), de-volatilization (pyrolysis zone), combustion (oxidation zone), gasification, and water-gas shift reaction (reduction zone). However, there is no sharp margin between these reactions (Okolie et al., 2019).

Dehydration or drying is the first step of the gasification process, which occurs at around 100°C. At this stage, the loss of water from the reacting molecule or ion reduces the water or moisture content of the feedstock. At this stage, about 10–20% of weight loss occurs. The water vapor/moisture flows downwards of the gasifier and is mixed with the water vapor generated in the oxidation zone and reduced to hydrogen. Then de-volatilization of feed materials takes place in the temperature range between 300°C and 500°C in the absence of oxygen. The de-volatilization process is also known as pyrolysis. The pyrolysis of carbonaceous material results in solid, liquid, and gaseous products depending on various parameters such as the heating rate, temperature, types of reactor and feedstock composition (Qiang et al., 2009; Puig-Arnavat and Coronas, 2010; Shadangi and Mohanty, 2014; Nanda et al., 2016). The solid products are generally pyrolytic char, whereas pyrolytic liquid includes tars, oil, heavy hydrocarbons and water. The pyrolytic gas is a mixture of CO_2, H_2O, CO, C_2H_2, C_2H_4, C_2H_6, and C_{2+} gasses. The longer residence time of the pyrolytic vapor in the reactor favors the conversion to char and gas where faster heating rate and shorter residence time favors conversion to oil at a moderate temperature.

Pyrolysis at a higher heating rate and higher temperature results in bio-oil and gaseous products. Pyrolysis of biomass with higher lignin content produces more char yield. However, the oil and gas yield increases in presence of high cellulose and hemicellulose content (Belgiorno et al., 2003; Couhert et al., 2009; Shadangi and Mohanty, 2014; Nanda et al., 2014). During pyrolysis breaking of weaker chemical bonds and de-volatilization releases, some condensable and non-condensable gasses along with high molecular weight char (Shadangi and Mohanty, 2014). The pyrolytic products transfer to the bottom of the gasifier, the hotter zone and known as the oxidation zone. Some of the products burnt in this zone and longer the residence time of the rest of the products convert to H_2, CH_4, CO, ethane, ethylene, etc. On combustion, the pyrolytic products react with oxygen and form CO_2 and CO. The heat generated by the process of combustion helps in the gasification reaction. At the oxidation zone, a sharp increase in temperature is observed (Kumar et al., 2009). This is an exothermic reaction.

The hot gas and char produced in the oxidation zone travel to the reduction zone where the sensible heat of the gas and char convert to the chemical energy of the producer gas. At this stage, the pyrolytic char reacts with CO_2 and steam to yield CO, H_2 and CH_4. The last step of the gasification reaction is the water-gas shift reaction followed by the methanation reaction. In this case, CO reacts with water and forms H_2. The generated H_2 gas reacts with CO and generates CH_4 and water. At the end of the process, ash is removed from the bottom of the gasifier. The important gasification reactions are stated as follows (Kumar et al., 2009; Puig-Arnavat and Coronas, 2010; Block et al., 2019; Shahabuddin et al., 2019). The combined pyrolysis and reforming is a worthy process to convert the mixture of biomass and waste plastic streams into H_2 (Shahabuddin et al., 2019).

Gasification with oxygen (partial oxidation):

$$C + 0.5O_2 \leftrightarrow CO \qquad (\Delta H^0_{298} = -111 \text{ kJ/mol}) \tag{4.1}$$

Combustion with oxygen (complete oxidation):

$$CO + O_2 \rightarrow CO_2 \qquad (\Delta H^0_{298} = -283 \text{ kJ/mol}) \tag{4.2}$$

Gasification with CO_2 (Boudouard reaction):

$$C + CO_2 \leftrightarrow 2CO \qquad (\Delta H^0_{298} = +173 \text{ kJ/mol}) \qquad (4.3)$$

Gasification with steam (Water-gas shift reaction):

$$CO + H_2O \leftrightarrow CO_2 + H_2 \qquad (\Delta H^0_{298} = -42 \text{ kJ/mol}) \qquad (4.4)$$

Gasification with H_2 (methanation):

$$C + 2H_2 \leftrightarrow CH_4 \qquad (\Delta H^0_{298} = -75 \text{ kJ/mol}) \qquad (4.5)$$

Steam reforming:

$$CH_4 + H_2O \rightarrow CO + 3H_2 \qquad (\Delta H^0_{298} = 206 \text{ kJ/mol}) \qquad (4.6)$$

4.5 CO-GASIFICATION OF BIOMASS AND WASTE PLASTICS

Co-gasification is the process where the mixture of biomass and plastics at a specific ratio used as feed for gasifier for the production of syngas. The ratio of biomass-to-plastics is one of the important factors directly related to the composition of the syngas from the co-gasification process. The co-gasification enhances the composition of H_2 to CO ratio in the syngas (Ruoppolo et al., 2012). The co-gasification helps in the use of biomass and plastics as per their availability (Brachi et al., 2014). In general, all types of biomass and plastics can be used simultaneously as feed in this process. Since plastic is a polymer and composition of hydrocarbon, its use with biomass as the feed for gasification increases the yield of gas and decreases the formation of char and tar (Lopez et al., 2015; Moghadam et al., 2014).

It was reported that the co-gasification of plastics and biomass minimizes the problem created during the operation of pyro-gasification of plastics only (Pinto et al., 2002; Ruoppolo et al., 2012). Concerning this, Brachi et al. (2014) stated that the co-gasification of biomass and polymer blended samples such as PET and tar reduced the problem created by gasification of only PET and tar. During the gasification of plastic material only, with the increase in the temperature, the plastics become viscous and sticky and results in more char and tar compared to gas yield. However, the addition of low-density wood or biomass reduces such types of problems. Similarly, Robinson et al. (2016) reported that the use of composite wood/PET pellets at a 1:1 ratio prevented the formation of coke in a fluidized bed gasifier. It was also mentioned that the utilization of pellets (biomass and PET) enhanced the fluidization capacity and heat transfer as well due to the density difference and hence the coke formation reduced. Simultaneously the additives used for the preparation of pellets such as organic or inorganic compounds may act as a catalyst during gasification. The formation of several pollutants is also possible due to the presence of additives. The additives may also act as a catalyst during the gasification process.

4.5.1 *EFFECTS OF PLASTIC-TO-BIOMASS RATIO ON CO-GASIFICATION*

Co-gasification experiments using biomass and various plastics are studied by several researchers and reported the effect of co-gasification on syngas composition. It was reported that the addition of 20% polystyrene with biomass increased the yield of liquid and lowered the yield of gas. However, the concentration of H_2, CO, and CO_2 were higher in the gas compared to CH_4 and C_nH_m. The research also confirmed that the yield of gas was higher when 20% of polyolefin was mixed with biomass compared to polystyrene (Alvarez et al., 2014; Wu and Williams, 2010).

Ephraim et al. (2016) reported the increase in the concentration of H_2 in the gas with the addition of 30% polystyrene with biomass at 750°C under N_2 atmosphere. The concentration of H_2 was quite more compared to CH_4, C_nH_m, CO, and CO_2. The presence of the aromatic cycle in the polystyrene, steam gasification at a higher temperature at about 800°C may increase the yield of H_2 gas is reported by Alvarez et al. (2014). The catalytic co-gasification of wood sawdust and 20% PP, HDPE, PS, and polyolefines using with and without Ni-based catalyst in a two-stage fixed-bed steam gasification setup at 800°C confirmed that co-gasification with polyolefines yielded a higher amount of gas and a lower amount of char and tar. A higher amount of polyolefines in biomass and plastic mixture produced a higher concentration of H_2, C_1–C_4 hydrocarbons and a lower concentration of CO and CO_2. It was also stated that the blending of PP with woodchip from 0–20% the yield of gas increased from 51.6% to 57% in which the concentration of H_2 also increased from 30.3% to 36.1% without the catalyst. When the same experiment was performed using a Ni-based catalyst, the yield of H_2 was elevated to 52.1% (Alvarez et al., 2014).

The steam gasification of municipal solid waste (MSW), rubber, plastic, and wood were investigated at 700°C using steam in a fixed bed gasifier by Lee et al. (2014). The research concluded that gasification at higher temperature steam was favorable for the production of H_2 rich (50–60%), gaseous fuel including 10% CO and CO_2 and about 3% CH_4 due to water gas sifting reaction. However, the concentration of H_2 in the product gas was higher for plastics compared to MSW and wood only. Simultaneously, the yield of CO, CO_2 and CH_4 in the product gas was lower when plastic was mixed with biomass. It was also investigated that gasification of individual wood and MSW yielded low H_2 and higher CO concentration in the product gas (Lee et al., 2014).

Ephraim et al. (2016) reported that the increase in the amount of PVC in the mixture of biomass and PVC increased the yield of gas including a higher concentration of H_2 and HCl, whereas the concentration of CH_4 and C_nH_m in the gas remained constant. The study also confirmed that the yield of tar was decreased with a higher concentration of PVC in the mixture. However, a variation was noticed in the yield of char. The char yield was increased up to 30% of PVC and later on, decreased with an increase in the amount of PVC.

Brachi et al. (2014) and Robinson et al. (2016) reported that PET is not a suitable plastic for co-gasification with wood to produce H_2. The addition of PET with wood enhanced the yield of Tar, lowered the H_2 concentration and increased the concentration of CO_2 and CO as well compared to wood. However, the yield of H_2 increased up to 50% with increasing the amounts of PE in the feed during steam co-gasification of biomass and Pinto et al. (2002) observed PE at 835°C using circular in cross-section gasifier. The concentration of H_2 in the gas remained constant beyond the 20% PET in the feed mixture. The reduction in the concentration of CO and CO_2 in the product gas with increasing PE concentration was also reported. Ahmed et al. (2011) reported that the optimum concentration of PE for gasification of PE/woodchip mixtures was about 80% for a higher yield of H_2.

Lopez et al. (2015) studied steam gasification of biomass and HDPE using a spouted bed reactor at 900°C. The study reported that the use of 2.5 times HDPE than biomass in the feed enhanced the yield of syngas and reduced the yield of tar and char. However, the concentration of H_2 was higher at about 57% at 50% HDPE in the feed. Further increase in the concentration of HDPE in feed reduced the concentration of H_2, CO, and CO_2. The effect of moisture content (from 9.5% to 27%) in the biomass samples in the mixture of rubberwood chips and rubber waste on the yield of H_2 was investigated at 800°C and reported that the mixture of rubberwood chips (27% moisture content) and 20% waste rubber increased yield of H_2 and high calorific value syngas compared to air-dried biomass (Kaewluan and Pipatmanomai, 2011). This confirmed that the utilization of moist biomass rather than air-dried biomass is favorable for H_2 production. Tavares et al. (2018) experimented using several ratios of PET to biomass (from 0% to 100%) at 600°C and reported that rising the PET ratio, H_2 concentration was enhanced from 16% (100% biomass) to 33% (0% biomass).

4.5.2 EFFECTS OF STEAM-TO-BIOMASS RATIO OR STEAM-TO-FUEL RATIO ON CO-GASIFICATION

Steam-to-biomass (S/B) ratio or steam-to-fuel (S/F) ratio is defined as the rate of introduction of steam to the rate of introduction of feed (biomass) to the gasifier (Tavares et al., 2018). S/B ratio or steam to fuel ratio is one of the important parameters, which directly affects the H_2 yield. The introduction of steam increases the partial pressure inside the gasified and hence enhances the H_2 concentration in the syngas. Pinto et al. (2002) reported that gasification of 40% PE and 60% biomass mixture at 835°C at a maximum S/F ratio of 0.75 resulted in a higher concentration of H_2 in the syngas. At this ratio, the decrease in the yield of CO, CH_4 and C_nH_m and increase in the yield of CO_2 were also noticed.

Tavares et al. (2018) conducted a gasification experiment by using 50% PET + 50% biomass and 90% PET+10% biomass at 800°C by varying the S/B from 0.8 to 2 and obtained a similar profile for each gas component. The effect of S/B ratio the yield visualized that the yield of H_2 and CO_2 increased with increasing the S/B ratio and the highest yield was perceived about 58% at 800°C at the S/B ratio of 2. The study also reported that the yield of CO increased and CO_2 decreased by increasing the concentration of PET in the feed.

4.5.3 EFFECTS OF EQUIVALENCE RATIO ON CO-GASIFICATION

Equivalence ratio (ER) is defined as the weight of air to the weight of biomass, relative to the stoichiometric weight air to stoichiometric weight biomass required to complete the reaction as shown below (Tavares et al., 2018).

$$ER = (O_2/\text{biomass})_{\text{weight}} \,/\, (O_2/\text{biomass})_{\text{stoichiometric}} \quad (4.7)$$

The ER ratio directly affects the composition and production of syngas and tar. A higher ER ratio means available for more oxygen during gasification. This enhances the formation of more H_2O, CO, CO_2 and tar, and decreases the yield of H_2 when oxygen reacts with hydrocarbons (Aznar et al., 2006). Tavares et al. (2018) reported that increasing the ER value reduced the concentration of H_2, CO, and CO_2 in the syngas. Since lower ER value causes incomplete gasification, a higher ER value increases the combustion and the concentration of CO_2 in the syngas along with char formation. This reduces the efficiency of a gasifier. Hence, it can be said that oxidation reactions take place once the actual oxygen to biomass ratio approaches the stoichiometric amount and results in CO_2 and H_2O rather than other combustible gasses. Thus, it can be considered that air is not a good gasifying agent for the gasifier.

Pinto et al. (2003) conducted co-gasification using coal, biomass, and, waste plastics in presence of air and steam mixture in a fluidized bed reactor by varying the oxygen/fuel ratio and obtained that the partial combustion reactions were favored when the oxygen/fuel ratio increased and resulted lower the concentration of H_2, CH_4 and other hydrocarbon and higher CO, CO_2 in the syngas. Gasification at a higher oxygen/fuel ratio enhances the yield of product gas but reduces the heating value since the product gas is rich in CO_2. Mastellone et al. (2012) also stated similar observations.

4.5.4 EFFECTS OF GASIFYING AGENT ON CO-GASIFICATION

A gasifying agent plays an important role in the yield and composition of syngas. Some of the gasifying agents are oxygen, air, CO_2, N_2, steam, and a mixture of any gasses. Gasification is an endothermic process, and hence, heat must be supplied during the process. Since the cost is also an important factor, the air is used as a gasifying agent. Ephraim et al. (2016) reported that the gasifying agent intensely affects the reaction rate at the entrance zone of the gasifier. If the air is used as a gasifying agent, the possibility of formation of CO_2 and H_2O is more and the yield of H_2 is less (about 7–12%), which produce low heating value syngas (< 6 MJ/m^3) due to the presence of N_2 in the gas (Aznar et al., 2006). The use of pure O_2 improves the calorific value syngas (10–20 MJ/Nm^3) and increase the cost as well. However, O_2 can mix with steam to improve the yield and H_2 concentration in the syngas and heating value (12–14 MJ/m^3) where the requirement of external heat is less compared to only steam.

The yield of H_2 concentration is higher and about 50–55% if only steam is used as a gasifying agent, whereas it has one drawback such as the yield of tar is more. Hence, the use of pure steam increases the yield of H_2 and reduces the concentration of CO and CO_2. The use of co-gasifying agent O_2 with steam decreases the formation of char and tar. Hence, O_2 can mix with steam to produce higher heating value syngas with good (H_2/CO ratio) (Ephraim et al., 2016). However, increasing the concentration of O_2 in the mixture of air and steam increases the partial combustion reaction and as a result, the syngas is diluted with N_2, increased the CO, CO_2 concentration, and hence decreases the heating value (Pinto et al., 2003).

Lee et al. (2014) reported that the lower heating value of the syngas produced by using steam is twice than air blown gasifier because of extra water-gas shift reaction occurs in steam gasification. Ephraim et al. (2016) confirmed that compared to the N_2 gasifying agent, the yield of syngas is better when steam and CO_2 (dry gasification) are used. This happens may be due to the Boudouard and water-gas shift reaction takes place with char (Ahmed and Gupta, 2011). However, Tavares et al. (2018) confirmed that the O_2 proved better as a gasifying agent compared to air. Co-gasification of paper and polystyrene in the presence of CO_2 as a gasifying agent enhanced the yield of CO and decreased the yield of H_2 (Déparrois et al., 2019). Therefore, the selection of gasifying agents and maintaining proper oxygen-to-steam ratio for gasification is most important.

4.5.5 EFFECTS OF TEMPERATURE ON CO-GASIFICATION

The temperature has a substantial effect on the yield of H_2 during the co-gasification of biomass and plastics. The syngas composition highly depends on the operating temperature. The carbon conversion rate increases with the increase in temperature. Higher temperature gasification decreases the yield of tar, increases the yield of syngas with a higher concentration of H_2, CO, and lowers the concentration of CO_2, N_2, CH_4, and other hydrocarbon gases. It was also reported that the yield of H_2 remained constant when 50% of PET and 50% of biomass were co-gasification at 900°C to 1200°C (Tavares et al., 2018). Similarly, fluidized bed co-gasification of pinewood sawdust and polyethylene (PE) produced a higher concentration of H_2 in the syngas at 830°C and H_2 concentration went on increasing with increasing PE content (from 10% to 60%) in the feed. It was also reported that the concentration of CO_2 increased and CO decreased with rising the PE content in the feed above 830°C (Pinto et al., 2002).

The effect of gasification temperature on H_2 yield was also studied by Pinto et al. (2003) using coal, biomass, and plastics wastes in a fixed bed reactor in the presence of air/steam mixture in the temperature range of 750–890°C. The study concluded that feed containing coal+pine+PE (60% + 20% + 20%) yielded higher H_2 content compared to other ratios. The yield of H_2 was increased by 70%, CH_4 and other hydrocarbon decreased by 30–63% when the temperature amplified from 750°C to 890°C. The formation of H_2 with temperature follows a linear trend. However, the rise and fall in the concentration of CO, CO_2 and CH_4 concerning temperature are not linear. It was reported that the concentration of CO first decreases and then increases above 830°C, CO_2 increases and decreases beyond 830°C (Aznar et al., 2006; Pinto et al., 2003). Moreover, the yield not only depends on the temperature but also other factors that are discussed here.

4.5.6 EFFECTS OF CATALYSTS ON CO-GASIFICATION

Catalyst is used to enhance the rate of reaction that boosts the formation of the H_2. It is a vital factor of the co-gasification process. Several research studies conducted on gasification using Ni-based catalysts reported that during the steam gasification, Ni-based catalysts enhance the rate of water-gas shift reaction and result in a high rate of H_2 formation by rupturing the C-C bond and by cracking and reforming reactions of pyrolytic products. The deposition of char on the active sites of the Ni catalyst is the cause of the catalyst deactivation. The yield of H_2 increases with the increase in the amount of Ni catalysts (Anis and Zainal, 2011; Wu et al., 2011, Basagiannis, and Verykios, 2007).

Song et al. (2010) reported that the H_2 production efficiency of the Ni catalyst was better compared to Ni, NiO, and Mg. The literature revealed that the catalyst support also plays an essential role. Several researchers mention the importance of supported Ni-based catalysts on the co-gasification (Alvarez et al., 2014; Ruoppolo et al., 2012; Brachi et al., 2014). Kumagai et al. (2015) synthesized Ni-Mg-Al-Ca catalysts with a varying concentration of Ca and studied the co-gasification of wood sawdust/ PP mixture. The study reported that the Ni/Mg/Al/Ca catalyst at 1:1:1:4 ratio yielded the highest 39.6 mol H_2/g Ni. The adsorption of CO_2 by CaO reduced the yield of CO_2 and enhanced the H_2 yield.

Alvarez et al. (2014) used Ni/Al_2O_3 catalyst for co-pyrolysis of wood sawdust and plastic wastes (20%). The study reported that catalyst enhanced the yield of H_2 about 36.1 vol% in syngas. The yield of syngas was also higher about 56.9 wt.%. Zhang et al. (2019) observed that the bimetallic catalyst such as Ni@CNF/PCs yielded the highest H_2 and CO from co-gasification of biomass/plastic (1:2 ratio) at 700°C compared to monometallic catalyst. The yield of H_2 increased from 2.23 mmol/g to 24.73 mmol/g and the concentration of CO increased from 3.20 mmol/g to 6.79 mmol/g without the catalyst to with catalyst. However, Pinto et al. (2019) reported that both dolomite and Ni-based catalysts increased the efficiency of H_2 production where the complete reduction in the tar yield was noticed for Ni-based catalytic reaction.

4.6 CONCLUSIONS

Co-gasification of biomass and plastic mixture has several advantages over individual gasification. It is a good way of minimization of wastes. This process can use all types of biomass and waste plastics to produce H_2. Not only H_2 energy, CO, and many types of low carbon number hydrocarbon fuel can be produced to use a fuel. Co-gasification of biomass and waste plastics decreases the problems that arise during the process. The utilization of low-cost catalyst enhances the yield and increases the efficiency of H_2 production.

This review concluded that a co-gasification process is a good option for the production of H_2 energy. However, the design of the gasifier, biomass to plastic ratio, the flow rate of gasifying agents, and the catalyst types are the most important factors, which have to take into consideration while designing a gasification plant. It was observed that the yield of H_2 gas also depends on the types of plastics and their ratio in the feed. The co-gasification behavior of biomass and PVC resulted in HCl as one of the products, which is very dangerous. Therefore, proper precautions have to be taken during the process. In general, the utilization of plastics enhances the yield of syngas and decreases the yield of tar and char. Co-gasification with PE, PP, and PS enhanced the yield of H_2, CH_4, hydrocarbons, and CO whereas decreased the yield of CO_2. Catalytic gasification can be a good option (especially Ni-based catalysts) to enhance the yield of H_2 and other combustible gasses and to reduce the formation of CO_2 in the syngas. Compared to other gasifying agents, the use of steam is better as it forms more hydrogen by water-gas shift reaction and resulting in fewer yields of char and tar.

KEYWORDS

- **biomass**
- **co-gasification**
- **high-density polyethylene**
- **hydrogen**
- **low-density polyethylene**
- **polyethylene**
- **polyethylene terephthalate**
- **polypropylene**
- **polystyrene**
- **polyurethane**
- **polyvinylchloride**
- **waste plastics**

REFERENCES

Ahmed, I. I., & Gupta, A. K., (2011). Kinetics of woodchips char gasification with steam and carbon dioxide. *Appl. Energy, 88*, 1613–1619.

Ahmed, I. I., Nipattummakul, N., & Gupta, A. K., (2011). Characteristics of syngas from co-gasification of polyethylene and woodchips. *Appl. Energy, 88*, 165–174.

Ahrenfeldt, J., Egsgaard, H., Stelte, W., Thomsen, T., & Henriksen, U. B., (2013). The influence of partial oxidation mechanisms on tar destruction in two stage biomass gasification. *Fuel, 112*, 662–680.

Alvarez, J., Kumagai, S., Wu, C., Yoshioka, T., Bilbao, J., Olazar, M., & Williams, P.T., (2014). Hydrogen production from biomass and plastic mixtures by pyrolysis-gasification. *Int. J. Hydrogen Energy, 39*, 10883–10891.

Anis, S., & Zainal, Z. A., (2011). Tar reduction in biomass producer gas via mechanical, catalytic and thermal methods: A review. *Renew. Sust. Energy Rev., 15*, 2355–2377.

Aznar, M. P., Caballero, M. A., Sancho, J. A., & Francés, E., (2006). Plastic waste elimination by co-gasification with coal and biomass in fluidized bed with air in pilot plant. *Fuel Process. Technol., 87*, 409–420.

Basagiannis, A. C., & Verykios, X. E., (2007). Catalytic steam reforming of acetic acid for hydrogen production. *Int J Hydrogen Energy, 32*, 3343–3355.

Belgiorno, V., Feo, G. D., Rocca, C. D., & Napoli, R. M. A., (2003). Energy from gasification of solid wastes. *Waste Manag., 23*, 1–15.

Block, C., Ephraim, A., Weiss-Hortala, E., Minh, D.P., Nzihou, A., & Vandecasteele, C., (2019). Co-pyro gasification of plastics and biomass, a review. *Waste Biomass Valor., 10*, 483–509.

Brachi, P., Chirone, R., Miccio, F., Miccio, M., Picarelli, A., & Ruoppolo, G., (2014). Fluidized bed co-gasification of biomass and polymeric wastes for a flexible end-use of the syngas: Focus on bio-methanol. *Fuel, 128*, 88–98.

Bridgwater, A. V., (2003). Renewable fuels and chemicals by thermal processing of biomass. *Chem. Eng. J., 91*, 87–102.

Chunfei, W., & Paul, T. W., (2010). Pyrolysis-gasification of plastics, mixed plastics and real world plastic waste with and without Ni-Mg-Al catalyst. *Fuel, 89*, 3022–3032.

Couhert, C., Commandre, J. M., & Salvador, S., (2009). Is it possible to predict gas yields of any biomass after rapid pyrolysis at high temperature from its composition in cellulose, hemicellulose and lignin? *Fuel, 88*, 408–417.

Déparrois, N., Singh, P., Burra, K. G., & Gupta, A. K., (2019). Syngas production from co-pyrolysis and co-gasification of polystyrene and paper with CO_2. *Appl. Energy, 246*, 1–10.

Ephraim, A., Pozzobon, V., Louisnard, O., Minh, D. P., Nzihou, A., & Sharrock, P., (2016). Simulation of biomass char gasification in a downdraft reactor for syngas production. *AIChE J., 62*, 1079–1091.

Kaewluan, S., & Pipatmanomai, S., (2011). Gasification of high moisture rubber woodchip with rubber waste in a bubbling fluidized bed. *Fuel Process. Technol., 92*, 671–677.

Kumagai, S., Alvarez, J., Blanco, P. H., Wu, C., Yoshioka, T., Olazar, M., & Williams, P. T., (2015). Novel Ni-Mg-Al-Ca catalyst for enhanced hydrogen production for the pyrolysis-gasification of a biomass/plastic mixture. *J. Anal. Appl. Pyrolysis, 113*, 15–21.

Kumar, A., Jones, D. D., & Hanna, M. A., (2009). Thermochemical biomass gasification: A review of the current status of the technology. *Energies, 2*, 556–581.

Lee, U., Chung, J. N., & Ingley, H. A., (2014). High-temperature steam gasification of municipal solid waste, rubber, plastic and wood. *Energy Fuels, 28*, 4573–4587.

Lopez, G., Erkiaga, A., Amutio, M., Bilbao, J., & Olazar, M., (2015). Effect of polyethylene co-feeding in the steam gasification of biomass in a conical spouted bed reactor. *Fuel, 153*, 393–401.

Mastellone, M. L., Zaccariello, L., Santoro, D., & Arena, U., (2012). The O_2-enriched air gasification of coal, plastics, and wood in a fluidized bed reactor. *Waste Manag., 32*, 733–742.

Moghadam, R. A., Yusup, S., Uemura, Y., Chin, B. L. F., Lam, H. L., & Al Shoaibi, A., (2014). Syngas production from palm kernel shell and polyethylene waste blend in fluidized bed catalytic steam co-gasification process. *Energy, 75*, 40–44.

Nanda, S., & Berruti, F., (2021a). A technical review of bioenergy and resource recovery from municipal solid waste. *J. Hazard. Mater., 403*, 123970.

Nanda, S., & Berruti, F., (2021b). Thermochemical conversion of plastic waste to fuels: A review. *Environ. Chem. Lett., 19*, 123–148.

Nanda, S., Dalai, A. K., Berruti, F., & Kozinski, J. A., (2016). Biochar as an exceptional bioresource for energy, agronomy, carbon sequestration, activated carbon and specialty materials. *Waste Biomass Valor., 7*, 201–235.

Nanda, S., Mohammad, J., Reddy, S. N., Kozinski, J. A., & Dalai, A. K., (2014). Pathways of lignocellulosic biomass conversion to renewable fuels. *Biomass Conv. Bioref., 4*, 157–191.

Nanda, S., Rana, R., Zheng, Y., Kozinski, J. A., & Dalai, A. K., (2017). Insights on pathways for hydrogen generation from ethanol. *Sustain. Energy Fuels, 1*, 1232–1245.

Okolie, J. A., Nanda, S., Dalai, A. K., & Kozinski, J. A., (2021). Chemistry and specialty industrial applications of lignocellulosic biomass. *Waste Biomass Valor, 12*, 2145–2169.

Okolie, J. A., Rana, R., Nanda, S., Dalai, A. K., & Kozinski, J. A., (2019). Supercritical water gasification of biomass: A state-of-the-art review of process parameters, reaction mechanisms and catalysis. *Sustain. Energy Fuels, 3*, 578–598.

Parthasarathy, P., & Narayanan, K. S., (2014). Hydrogen production from steam gasification of biomass: Influence of process parameters on hydrogen yield: A review. *Renew. Energy, 66*, 570–579.

Pinto, F., André, R. N., Franco, C., Lopes, H., Gulyurtlu, I., & Cabrita, I., (2009). Co-gasification of coal and wastes in a pilot-scale installation 1: Effect of catalysts in syngas treatment to achieve tar abatement. *Fuel, 88*, 2392–2402.

Pinto, F., Franco, C., André, R. N., Miranda, M., Gulyurtlu, I., & Cabrita, I., (2002). Co-gasification study of biomass mixed with plastic wastes. *Fuel, 81*, 291–297.

Pinto, F., Franco, C., André, R. N., Tavares, C., Dias, M., Gulyurtlu, I., & Cabrita, I., (2003). Effect of experimental conditions on co-gasification of coal, biomass and plastics wastes with air/steam mixtures in a fluidized bed system. *Fuel, 82*, 1967–1976.

Puig-Arnavat, M., Bruno, J. C., & Coronas, A., (2010). Review and analysis of biomass gasification models. *Renew. Sust. Energy Rev., 14*, 2841–2851.

Qiang, L., Wen-Zhi, L., & Xi-Feng, Z., (2009). Overview of fuel properties of biomass fast pyrolysis oils. *Energy Convers. Manag., 50*, 1376–1383.

Robinson, T., Bronson, B., Gogolek, P., & Mehrani, P., (2016). Comparison of the air-blown bubbling fluidized bed gasification of wood and wood-PET pellets. *Fuel, 178*, 263–271.

Ruoppolo, G., Ammendola, P., Chirone, R., & Miccio, F., (2012). H_2-rich syngas production by fluidized bed gasification of biomass and plastic fuel. *Waste Manag., 32*, 724–732.

Sarangi, P. K., & Nanda, S., (2020). Biohydrogen production through dark fermentation. *Chem. Eng. Technol., 43*, 601–612.

Shadangi, K. P., & Mohanty, K., (2013). Characterization of nonconventional oil containing seeds towards the production of bio-fuel. *J. Renew. Sustain. Energy, 5*, 033111–033113.

Shadangi, K. P., & Mohanty, K., (2014). Kinetic study and thermal analysis of the pyrolysis of non-edible oilseed powders by thermogravimetric and differential scanning calorimetric analysis. *Renew. Energy, 63*, 337–344.

Shahabuddin, M., Krishna, B. B., Bhaskar, T., & Perkins, G., (2020). Advances in the thermo-chemical production of hydrogen from biomass and residual wastes: Summary of recent techno-economic analyses. *Bioresour. Technol., 299*, 122557.

Song, H. J., Lee, J., Gaur, A., Park, J. J., & Park, J. W., (2010). Production of gaseous fuel from refuse plastic fuel via co-pyrolysis using low-quality coal and catalytic steam gasification. *J. Mater. Cycles Waste Manag., 12*, 295–301.

Tavares, R., Ramos, A., & Rouboa, A., (2018). Microplastics thermal treatment by polyethylene terephthalate-biomass gasification. *Energy Convers. Manage., 162*, 118–131.

Wu, C., & Williams, P. T., (2010). Pyrolysis-gasification of plastics, mixed plastics and real world plastic waste with and without Ni-Mg-Al catalyst. *Fuel, 89*, 3022–3032.

Wu, C., Wang, L., Williams, P. T., Shi, J., & Huang, J., (2011). Hydrogen production from biomass gasification with Ni/MCM-41 catalyst: Influence of Ni content. *Appl. Catal. B: Environ.*, 108–109.

Zhang, S., Zhu, S., Zhang, H., Liu, X., & Xiong, Y., (2019). High quality H_2-rich syngas production from pyrolysis-gasification of biomass and plastic wastes by NieFe@ Nanofibers/Porous carbon catalyst. *Int. J. Hydrogen Energy, 44*, 26193–26203.

CHAPTER 5

An Overview of Fermentative Hydrogen Production Technologies

PRAKASH K. SARANGI,[1] SONIL NANDA,[2] AJAY K. DALAI,[2] and JANUSZ A. KOZINSKI[3]

[1]*Directorate of Research, Central Agricultural University, Imphal, Manipur, India*
E-mail: sarangi77@yahoo.co.in (Prakash K. Sarangi)

[2]*Department of Chemical and Biological Engineering, University of Saskatchewan, Saskatoon, Saskatchewan, Canada*

[3]*Faculty of Engineering, Lakehead University, Thunder Bay, Ontario, Canada*

ABSTRACT

Sustainable production of hydrogen is necessitated in the present day due to a wide range of applications providing energy and fuels in multi-dimensional sectors. Waste biomass in this regard acts as a valuable resource for the production of biohydrogen. Organic wastes having emerging perspectives with high-energy content and other fuel characteristics can be explored for hydrogen production through various sustainable routes. Fermentation processes such as photo-fermentation and dark fermentation are the key tools for biohydrogen production using organic biomass and microorganisms. The potential of fermentation processes for hydrogen production is summarized in this chapter.

5.1 INTRODUCTION

At present, fossil fuels are used as important sources for the production of energy and chemicals. Since the industrial revolution and the vast use of automobiles, fossil fuels create drastic environmental pollution, thereby urging researchers to opt for alternative energy sources with fewer carbon emissions. Biofuels from biomass are considered as the clean and green alternative for mitigating fossil fuel-related pollution and climate change (Sarangi et al., 2009a, b, 2010; Nanda et al., 2014). Numerous organic wastes generated by agricultural sectors, forestry ecosystems, energy crop systems, food processing, municipal solid waste, livestock farming, industrial activities can be used as low-cost substrates for biological and thermochemical conversion to produce biofuels and value-added biochemicals (Sarangi et al., 2017, 2019; Sarangi and Nanda, 2019a,b,c). A high proportion of these waste materials are carbohydrates and phenolic in nature. Lignocellulosic biomass is a renewable resource on earth and it has attracted continuing efforts to produce biofuels and biochemicals (Nanda et al., 2013; Sarangi et al., 2017; Sarangi and Nanda, 2018; Sarangi, 2019; Sarangi et al., 2020; Okolie et al., 2020, 2021). Waste-to-energy conversion acts as the increasingly popular alternative for the treatment of agricultural and industrial wastes towards economic and ecological benefits (Nanda and Berruti, 2021a,b,c).

Among various types of biofuels, biohydrogen has gained a promising alternative having a wide array of applications in various sectors being used as energy and fuel processing (Reddy et al., 2014; Nanda et al., 2017b). Hydrogen can be used as a next-generation energy carrier and vector with zero carbon emissions because its combustion produces heat and water (Okolie et al., 2019; Nanda et al., 2020; Pasin et al., 2020; Bhatia et al., 2020; Sarangi et al., 2020). The production of hydrogen as biofuel has gained much research among researchers due to its clean fuel characteristics. The environmentally friendly production and utilization of hydrogen can be considered as one of the United Nations Sustainable Development Goals, which targets for enhancing international cooperation for facilitating the access of clean energy research and technology along with promotional investment in eco-friendly energy infrastructures (United Nations, 2018; Bundhoo, 2019).

Hydrogen generation and utilization are gaining an increasing interest in the transportation sector as fuel cells in contrast to fossil fuels. Therefore, hydrogen is regarded as the fuel of the future. Hydrogen has a high-energy content of 120–142 MJ/kg, which is greater than that of several conventional hydrocarbon fuels (Nanda et al., 2017b). There is a need for sustainable production routes for hydrogen generation due to over-dependence on the reforming of natural gas and fossil fuels, which are highly polluting (Das and Veziroglu, 2001; Singh et al., 2018). Other methods such as electrolysis, photo-catalysis, thermochemical processes (e.g., gasification), photo-electrochemical processing, photochemical methods and fermentative processes are a few eco-friendly routes for hydrogen production (Nanda et al., 2017b; Nguyen et al., 2020).

Fossil fuel feedstocks are used as the resources for the production of hydrogen through gasification and steam reforming methods. In contrast, biological methods are focused on the production of hydrogen, mostly through dark fermentation and photo-fermentation (Sarangi and Nanda, 2020). Organic biomass having more biodegradability and low crystallinity can be used for the production of hydrogen through biological methods. Various production technologies from renewable sources are summarized in Figure 5.1. In this chapter, the generation of biohydrogen by fermentation processes are described along with their process conditions.

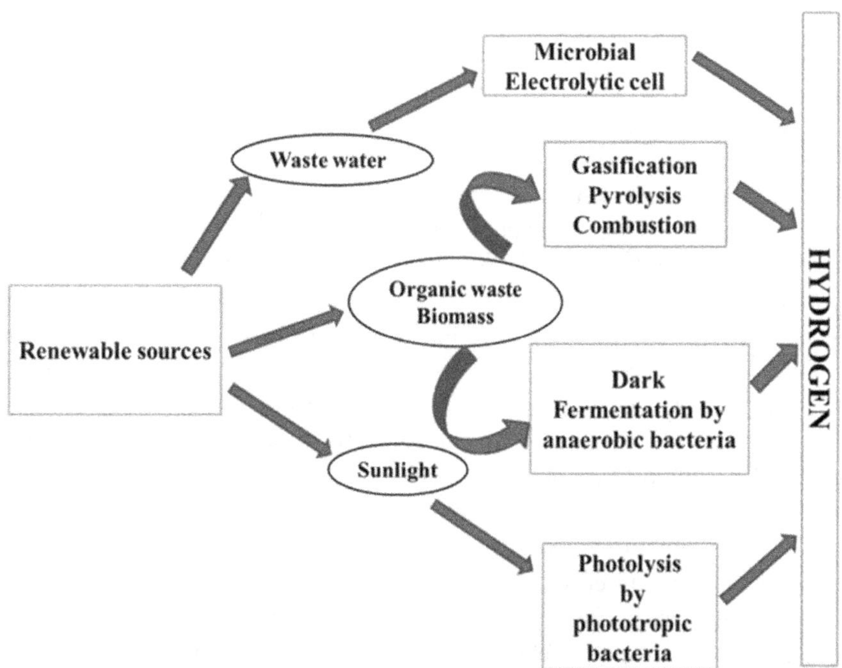

FIGURE 5.1 Production methods of biohydrogen from renewable sources.

5.2 BIOLOGICAL PRODUCTION OF HYDROGEN

The production of biohydrogen through the biological methods has been recognized as the most promising option for biohydrogen production due to fewer environmental risks, the negligible release of pollutants, relatively less energy input, low-cost, and mild operating conditions. The widely studies biological methods for hydrogen production are dark fermentation, light fermentation, and microbial electrolytic cells (Figure 5.1). Dark fermentation has great potential having a major share for efficient and economical production of hydrogen with higher rates and production (Sarangi and Nanda, 2020).

Biophotolysis of water is one of the alternatives for the production of biohydrogen by green algae and blue-green algae (cyanobacteria) directly or indirectly. In these processes, microorganism splits the water molecule into hydrogen ions and oxygen (Caudillo-Flores et al., 2017). The produced hydrogen ions are converted to hydrogen gas by the help of electrons contributed by reduced ferredoxin, which is supported by hydrogenase enzymes found in the microbial cells (Winkler et al., 2002). Cyanobacteria are photoautotrophic microorganisms that perform photosynthesis producing hydrogen in two stages (Levin et al., 2004). Various photosynthetic pigments such as chlorophyll, carotenoids, and phycobiliproteins are possessed by these cyanobacteria that support hydrogen production through the enzymes like hydrogenase and nitrogenase (Pinto et al., 2002). *Anabaena cylindrica* is a hydrogen-producing cyanobacterium, whereas *Anabaena variabilis* is hydrogen at high yields (Rupprecht et al., 2006). As the production of hydrogen from water takes place by two stages, it is regarded as the indirect bio-photolysis. Some nitrogen-fixing and non-nitrogen fixing hydrogen-producing cyanobacteria are employed in this process (Liu et al., 2006).

Microbial electrolysis cells are also very useful for the generation of biohydrogen by applying external electricity through bio-electrochemical reactions. Through microbial electrolysis cell (MEC), organic materials are oxidized by electrogenic bacteria in the anode chamber leading to the evolution of hydrogen in the cathode chamber. The process and configuration are equivalent to a microbial fuel cell in which electricity is generated. The complete mechanism works by the combination of bacterial electrolysis of organic matter and electrochemically assisted supplementation of external voltage towards the generation of hydrogen.

For enhanced production of biohydrogen, MECs can be combined with dark fermentation. This is because the thermodynamic limit of 4 mols of

H_2 per mole of glucose in the dark fermentation process can be combined for the utilization of the leftover energy from the substrate by MECs for supplementary hydrogen recovery (Figure 5.2). The conversion of complex organic materials into hydrogen during the dark fermentation process produces volatile fatty acids (VFA) by the help of acidogenic bacteria. The generation of VFAs can be used for the production of biohydrogen in MECs by electrogenic bacteria. Such an integration process improves the overall hydrogen yields attending high substrate degradation potential. This process of integration is very much favorable for the utilization of lignocellulosic biomass as the substrates. By single use of MFCs, the production rate and overall process efficiencies are low as compared to the integrated process methods. Furthermore, the scale of MECs follows the scale of dark fermentation, thereby achieving higher efficiencies (Lalaurette et al., 2009; Wang et al., 2011; Chookaew et al., 2014). A few bottlenecks have been detected during the move from dark fermentation to MECs like cell biomass, pH, and reactor designing, which overall reduces the efficiencies for hydrogen production.

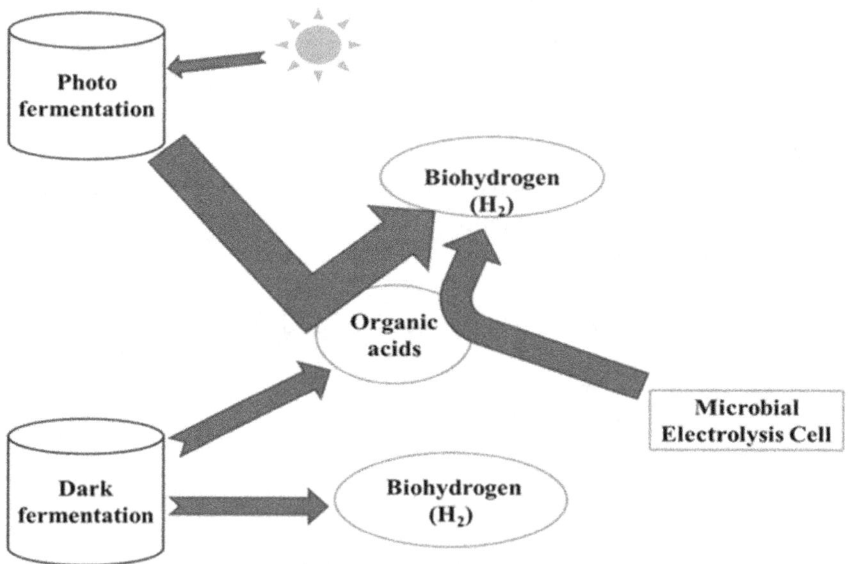

FIGURE 5.2 Combination of dark fermentation, photo-fermentation, and microbial electrolysis cell for biohydrogen production.

5.3 PHOTO-FERMENTATION

A special group of photoheterotrophic microorganisms such as purple non-sulfur bacteria mediates photo-fermentation (Wakerley et al., 2017). These organisms have great potential for the production of hydrogen using a wide array of organic and inorganic substances, which are an excellent source of electrons (Kuehnel and Reisner, 2018). These microorganisms can potentially convert organic wastes to hydrogen in the presence of light under nitrogen-deficit conditions. Because such bacteria lack Photosystem II, they help in the elimination of some difficulties associated with oxygen inhibition of hydrogen production. These microorganisms as compared to that of biophotolysis of water produce a substantial amount of hydrogen (Hao et al., 2018). In this process, biohydrogen production using photosynthetic bacteria is accomplished through a nitrogenase enzyme system in which light energy is necessary. The production of biohydrogen is detected through the conversion of reduced organic acids as the carbon source in the presence of light by purple non-sulfur bacteria with the help of the nitrogenase enzyme system (Sarangi and Nanda, 2020). In this process, purple non-sulfur photosynthetic bacteria perform photosynthesis in the absence of oxygen using light as an energy source for synthesizing hydrogen (Adessi and De Philippis, 2014; Eroglu and Melis, 2011). Bacterial photosystem produces two electrons with four adenosine triphosphate (ATP) molecules that help in the production of hydrogen from organic acids with a nitrogenase system. The purple non-sulfur bacteria, thereby increasing the biohydrogen recovery by integrating dark and photo-fermentation have detected the better use of organic acids as effluents from the dark fermentation (Liu et al., 2006).

Dark fermentation:

$$C_6H_{12}O_6 + 2H_2O \longrightarrow 2CH_3COOH + 2CO_2 + 4H_2 \qquad (5.1)$$

Photo-fermentation:

$$CH_3COOH + 2H_2O \xrightarrow{\text{Light energy}} 4H_2 + 2CO_2 \qquad (5.2)$$

Depending on the operation conditions and configuration of the bioreactors, dark fermentation also undergoes various pathways to produce different organic acids like lactate, propionate, and butyrate (Barbosa et al., 2001; Han et al., 2012):

$$\text{Lactate: } C_3H_6O_3 + 3H_2O \longrightarrow 6H_2 + 3CO_2 \quad (5.3)$$

$$\text{Propionate: } C_3H_6O_2 + 4H_2O \longrightarrow 7H_2 + 3CO_2 \quad (5.4)$$

$$\text{Butyrate: } C_4H_8O_2 + 6H_2O \longrightarrow 10H_2 + 4CO_2 \quad (5.5)$$

A wide array of purple non-sulfur species such as *Rhodopseudomonas palustris*, *Rhodospirillum rubrum*, *Rhodobacter capsulatus*, *Rhodobacter sphaeroides*, and *Rhodopseudomonas faecalis* have the potential for the conversion of organic acids into biohydrogen through the photo-fermentation process (Adessi and De Philippis, 2014; Eroglu and Melis, 2011). Some mixed cultures from wastewater sludge have been also found to produce biohydrogen through photo-fermentation (Mohan et al., 2008; Cheng et al., 2012; Xia et al., 2013).

5.4 DARK FERMENTATION

As far as other biological processes are concerned, dark fermentation is widely considered for relatively higher conversion rates and hydrogen yields. Biophotolysis of water and photo-fermentation require additional energy input. Conversely, dark fermentation does not need light energy, thereby involving moderate process conditions and less energy input (Garritano et al., 2017). Besides, the end products of dark fermentation are the volatile acids, which can be used as suitable feedstocks by methanogens for biomethane production. Lignocellulosic materials can be used for dark fermentation; however, for an efficient fermentation, the biomass might need pretreatment and saccharification to release fermentable sugars (Kucharska et al., 2015). The objective of pretreatment is to break down the cell walls of cellulosic biomass making the cellulose and hemicellulose more accessible to the hydrolysis process (Nanda et al., 2015).

In dark fermentation, the carbohydrate-rich feedstocks are converted to biohydrogen in the absence of oxygen and light by facultative and obligate anaerobic microorganisms. Hydrogen is generated by the activity of the hydrogenase enzyme (Das and Veziroglu, 2001; Li and Fang, 2007). This method works as opposed to aerobic respiration, which liberates oxygen and water as the end product (Das and Veziroglu, 2008; Wang and Wan, 2009).

The mechanism of dark fermentation starts from glucose as the initial substrate while subsequently producing pyruvate through glycolysis. Adenosine triphosphate (ATP) is generated from adenosine diphosphate

(ADP) and nicotinamide adenine dinucleotide (NADH) (Li and Fang, 2007). In the next step, pyruvate is converted to acetyl coenzyme A (acetyl-CoA) along with CO_2 and hydrogen by the catalytic action of pyruvate ferredoxin oxidoreductase and hydrogenase enzymes (Nanda et al., 2017a). According to the type of microorganisms and operating conditions, other products like formic acid may be produced from pyruvate, which is further transformed into H_2 and CO_2.

The conversion of acetyl-CoA is accomplished to acetate, ethanol, and butyrate (Li and Fang, 2007). Wide ranges of operational parameters that influence the dark fermentation are substrate type, temperature, loading rate, pH, reaction time, hydraulic retention time, hydrogen partial pressure, microorganism type, and bioreactor configurations, to name a few. Strictly anaerobic bacteria like *Clostridium*, methylotrophs, and methanogenic bacteria and facultative anaerobic bacteria like *Escherichia coli*, *Citrobacter*, and *Enterobacter* generally perform dark fermentation. Depending on the end product produced, the yield of biohydrogen is assumed. Acetate and butyrate are generally produced as fermentation products (Hawkes et al., 2007). The general biochemical reactions, which take place by facultative anaerobic microorganisms are presented below:

$$C_6H_{12}O_6 + 2H_2O \longrightarrow 2CH_3COOH + 2CO_2 + 4H_2 \tag{5.6}$$

$$C_6H_{12}O_6 \longrightarrow CH_3CH_2CH_2COOH + 2CO_2 + 2H_2 \tag{5.7}$$

Clostridium butyricum produces butyric acid and acetic acid along with hydrogen (Hawkes et al., 2007; Nanda et al., 2020). Another pathway also produces propionate using *Clostridium articum*, whereas no/little hydrogen is produced when ethanol and lactate are produced by *Clostridium barkeri* (Khanal et al., 2004).

$$C_6H_{12}O_6 + 2H_2 \longrightarrow 2CH_3CH_2COOH + 2H_2O \tag{5.8}$$

$$C_6H_{12}O_6 \longrightarrow 2CH_3CH_2OH + 2CO_2 \tag{5.9}$$

$$C_6H_{12}O_6 \longrightarrow CH_3CHOHCOOH + 2CO_2 \tag{5.10}$$

For mitigation of the environmental issues, the biochemical production of hydrogen has great perspectives. Biohydrogen production by dark fermentation compared to other techniques such as biophotolysis of water and photo-fermentation is advantageous due to the high yield of hydrogen. The nature of microorganisms and feedstocks used in dark fermentation play vital roles in the biological degradation of waste biomass to biohydrogen. The integration of dark fermentation, photo-fermentation, and electrochemical systems can be explored for biohydrogen generation in large-scale systems. Besides, exploration of dark fermentation with a two-stage digestion process in the absence of oxygen for the generation of biohythane (a hydrogen-methane blend) could be supportive in novel bioenergy utilization for the future energy crisis (Bolzonella et al., 2018).

According to the lifecycle assessment study conducted by Romagnoli et al. (2011), biohydrogen exhibits more advantages over fossil fuels towards electricity generation. Various biomasses are also compared for evaluating the energetic and environmental issues of hydrogen production (Djomo and Blumberga, 2011). Furthermore, the lifecycle assessment studies reveal that biohydrogen production pathways have a significant impact on the environment by using various biomasses and eco-friendly processes (Wulf et al., 2017). Hosseini et al. (2015) also supported the suitability of eco-friendly technologies for biohydrogen production by photosynthetic bacteria. Dadak et al. (2016) provided the basic tool of the exergy concept for photobiological hydrogen production. The yield of biohydrogen is dependent on the operational costs and production rate.

5.5 CONCLUSIONS

Hydrogen generation through the microbial conversion process has gained much interest due to the utilization of renewable feedstocks and environmentally friendly processes. Fermentative methods for hydrogen generation through dark fermentation and photo-fermentation along with optimization of various parameters are necessitated for maximum hydrogen recovery. The major pathways for biological hydrogen generation are photolysis of water, oxidation of organic acids by photo-fermentation and dark fermentation. Dark fermentation is one of the widely used processes of hydrogen production as it provides low input of energy from the feedstock. However, process parameters such as temperature, hydraulic retention time, the partial pressure of hydrogen, pH, reaction time, bioreactor type, substrate concentration, microorganism type, pretreatment process have key roles in regulating the hydrogen generation with maximum recovery through fermentative processes.

KEYWORDS

- **acetyl coenzyme A**
- **adenosine triphosphate**
- **biofuels**
- **biohydrogen**
- **dark fermentation**
- **microbial electrolysis cell**
- **nicotinamide adenine dinucleotide**
- **photo-fermentation**
- **volatile fatty acids**
- **waste biomass**

REFERENCES

Adessi, A., & De Philippis, R., (2014). Photobioreactor design and illumination systems for H_2 production with anoxygenic photosynthetic bacteria: A review. *Int. J. Hydrogen Energy, 39*, 3127–3141.

Barbosa, M. J., Rocha, J. M., Tramper, J., & Wijffels, R. H., *(2001).* Acetate as a carbon source for hydrogen production by photosynthetic bacteria. *J. Biotechnol., 85*, 25–33.

Bhatia, L., Sarangi, P. K., & Nanda, S., (2020). Current advancements in microbial fuel cell technologies. In: Nanda, S., Vo, D. V. N., & Sarangi, P. K., (eds.), *Biorefinery of Alternative Resources: Targeting Green Fuels and Platform Chemicals* (pp. 477–494). Springer, Singapore.

Bolzonella, D., Battista, F., Cavinato, C., Gottardo, M., Micolucci, F., Lyberatos, G., & Pavan, P., (2018). Recent developments in biohythane production from household food wastes: A review. *Bioresour. Technol., 257*, 311–319.

Bundhoo, Z. M. A., (2019). Potential of bio-hydrogen production from dark fermentation of crop residues: A review. *Int. J. Hydrogen Energy, 44*, 17346–17362.

Caudillo-Flores, U., Muñoz-Batista, M. J., Cortés, J. A., Fernández-García, M., & Kubacka, A., (2017). UV and visible light driven H_2 photo-production using Nb-doped TiO_2: Comparing Pt and Pd co-catalysts. *Mol. Catal., 437*, 1–10.

Cheng, J., Xia, A., Liu, Y., Lin, R., Zhou, J., & Cen, K., (2012). Combination of dark and photo-fermentation to improve hydrogen production from *Arthrospira platensis* wet biomass with ammonium removal by zeolite. *Int. J. Hydrogen Energy, 37*, 13330–13337.

Chookaew, T., Prasertsan, P., & Ren, Z. J., (2014). Two-stage conversion of crude glycerol to energy using dark fermentation linked with microbial fuel cell or microbial electrolysis cell. *New Biotechnol., 31*, 179–184.

Dadak, A., Aghbashlo, M., Tabatabaei, M., Najafpour, G., & Younesi, H., (2016). Exergy analysis as a tool for decision making on substrate concentration and light intensity in photobiological hydrogen production. *Energy Technol., 4*, 429–440.

Das, D., & Veziroglu, T. N., (2001). Hydrogen production by biological processes: A survey of literature. *Int. J. Hydrogen Energy, 26*, 13–28.

Das, D., & Veziroglu, T., (2008). Advances in biological hydrogen production processes. *Int. J. Hydrogen Energy, 33*, 6046–6057.

Djomo, S. N., & Blumberga, D., (2011). Comparative life cycle assessment of three biohydrogen pathways. *Bioresour. Technol., 102*, 2684–2694.

Eroglu, E., & Melis, A., *(2011)*. Photobiological hydrogen production: Recent advances and state-of-the-art. *Bioresour. Technol., 102,* 8403–8413.

Garritano, A. N., De Sá, L. V., Aguieiras, É. G., Freire, D. G., & Ferreira-Leitão, V. S., (2017). Efficient biohydrogen production via dark fermentation from hydrolyzed palm oil mill effluent by non-commercial enzyme preparation. *Int. J. Hydrogen Energy, 42*, 29166–29174.

Han, H., Liu, B., Yang, H., & Shen, J., (2012). Effect of carbon sources on the photobiological production of hydrogen using *Rhodobacter sphaeroides* RV. *Int. J. Hydrogen Energy, 37*, 12167–12174.

Hao, H., Zhang, L., Wang, W., & Zeng, S., (2018). Facile modification of TiO_2 with nickel sulfide and sulfate species for photo reformation of cellulose into H_2. *ChemSusChem, 11*, 2810–2817.

Hawkes, F., Hussy, I., Kyazze, G., Dinsdale, R., & Hawkes, D., (2007). Continuous dark fermentative hydrogen production by mesophilic microflora: Principles and progress. *Int. J. Hydrogen Energy, 32*, 172–184.

Hossein, S. S., Aghbashlo, M., Tabatabaei, M., Najafpour, G., & Younesi, H., (2015). Thermodynamic evaluation of a photobioreactor for hydrogen production from syngas via a locally isolated *Rhodopseudomonas palustris* PT. *Int. J. Hydrogen Energy, 36*, 7866–7871.

Khanal, S., Chen, W. H., Li, L., & Sung, S., (2004). Biological hydrogen production: Effects of pH and intermediate products. *Int. J. Hydrogen Energy, 29*, 1123–1131.

Kucharska, K., Łukajtis, R., Słupek, E., Cieśliński, H., Rybarczyk, P., & Kamiński, M., (2018). Hydrogen production from energy poplar preceded by MEA pre-treatment and enzymatic hydrolysis. *Molecules, 23*, 1–21.

Kuehnel, M. F., & Reisner, E., (2018). Solar hydrogen generation from lignocellulose. *Angew. Chem. Int. Ed., 57*, 3290–3296.

Lalaurette, E., Thammannagowda, S., Mohagheghi, A., Maness, P. C., & Logan, B. E., (2009). Hydrogen production from cellulose in a two-stage process combining fermentation and electro hydrogenesis. *Int. J. Hydrogen Energy, 34*, 6201–6210.

Levin, D. B., Pitt, L., & Love, M., (2004). Biohydrogen production: Prospects and limitations to practical application. *Int. J. Hydrogen Energy, 29*, 173–185.

Li, C., & Fang, H. H. P., (2007). Fermentative hydrogen production from wastewater and solid wastes by mixed cultures. *Crit. Rev. Environ. Sci. Technol., 37*, 1–39.

Liu, H., Hu, H., Chignell, J., & Fan, Y., (2010). Microbial electrolysis: Novel technology for hydrogen production from biomass. *Biofuels, 1*, 129–142.

Liu, J., Bukutin, V. E., & Tsygankov, A. A., (2006). Light energy conversion into H_2 by *Anabaena variables* mutant PK84 dense culture exposed in nitrogen limitations. *Int. J. Hydrogen Energy, 31*, 1591–1596.

Mohan, S. V., Srikanth, S., Dinakar, P., & Sarma, P. N., (2008). Photo-biological hydrogen production by the adopted mixed culture: Data enveloping analysis. *Int. J. Hydrogen Energy, 33*, 559–569.

Nanda, S., & Berruti, F., (2021a). A technical review of bioenergy and resource recovery from municipal solid waste. *J. Hazard. Mater., 403*, 123970.

Nanda, S., & Berruti, F., (2021b). Municipal solid waste management and landfilling technologies: A review. *Environ. Chem. Lett., 19*, 1433–1456.

Nanda, S., & Berruti, F., (2021c). Thermochemical conversion of plastic waste to fuels: A review. *Environ. Chem. Lett., 19*, 123–148.

Nanda, S., Li, K., Abatzoglou, N., Dalai, A. K., & Kozinski, J. A., (2017a). Advancements and confinements in hydrogen production technologies. In: Dalena, F., Basile, A., & Rossi, C., (eds.), *Bioenergy Systems for the Future* (pp. 373–418). Woodhead Publishing, Elsevier, UK.

Nanda, S., Maley, J., Kozinski, J. A., & Dalai, A. K., (2015). Physico-chemical evolution in lignocellulosic feedstocks during hydrothermal pretreatment and delignification. *J. Biobased Mater. Bioenergy, 9*, 295–308.

Nanda, S., Mohammad, J., Reddy, S. N., Kozinski, J. A., & Dalai, A. K., (2014). Pathways of lignocellulosic biomass conversion to renewable fuels. *Biomass Conv. Bioref., 4*, 157–191.

Nanda, S., Mohanty, P., Pant, K. K., Naik, S., Kozinski, J. A., & Dalai, A. K., (2013). Characterization of North American lignocellulosic biomass and biochars in terms of their candidacy for alternate renewable fuels. *Bioenergy Res., 6*, 663–677.

Nanda, S., Rana, R., Vo, D. V. N., Sarangi, P. K., Nguyen, T. D., Dalai, A. K., & Kozinski, J. A., (2020). A spotlight on butanol and propanol as next-generation synthetic fuels. In: Nanda, S., Vo, D. V. N., & Sarangi, P. K., (eds.), *Biorefinery of Alternative Resources: Targeting Green Fuels and Platform Chemicals* (pp. 105–126). Springer Nature, Singapore.

Nanda, S., Rana, R., Zheng, Y., Kozinski, J. A., & Dalai, A. K., (2017b). Insights on pathways for hydrogen generation from ethanol. *Sustain. Energy Fuels, 1*, 1232–1245.

Nguyen, D. T., Nguyen, V. H., Nanda, S., Vo, D. V. N., Nguyen, V. H., Tran, T. V., Nong, L. X., et al., (2020). $BiVO_4$ photocatalysis design and applications to hydrogen production, degradation of organic compounds and reduction of CO_2 to fuels: A review. *Environ. Chem. Lett., 18*, 1779–1801.

Okolie, J. A., Nanda, S., Dalai, A. K., & Kozinski, J. A., (2021). Chemistry and specialty industrial applications of lignocellulosic biomass. *Waste Biomass Valor, 12*, 2145–2169.

Okolie, J. A., Nanda, S., Dalai, A. K., Berruti, F., & Kozinski, J. A., (2020). A review on subcritical and supercritical water gasification of biogenic, polymeric and petroleum wastes to hydrogen-rich synthesis gas. *Renew. Sust. Energy Rev., 119*, 109546.

Okolie, J. A., Rana, R., Nanda, S., Dalai, A. K., & Kozinski, J. A., (2019). Supercritical water gasification of biomass: A state-of-the-art review of process parameters, reaction mechanisms and catalysis. *Sustain. Energy Fuels, 3*, 578–598.

Pasin, T. M., De Almeida, P. Z., De Almeida, S. A. S., Da Conceição, I. J., & De Teixeira De, M. P. M. L., (2020). Bioconversion of agro-industrial residues to second-generation bioethanol. In: Nanda, S., Vo, D. V. N., & Sarangi P. K., (eds.), *Biorefinery of Alternative Resources: Targeting Green Fuels and Platform Chemicals* (pp. 23–47). Springer, Singapore.

Pinto, F. A. L., Troshina, O., & Lindblad, P., (2002). A brief look at three decades of research on cyanobacterial hydrogen evolution. *Int. J. Hydrogen Energy, 27*, 1209–1215.

Reddy, S. N., Nanda, S., Dalai, A. K., & Kozinski, J. A., (2014). Supercritical water gasification of biomass for hydrogen production. *Int. J. Hydrogen Energy, 39*, 6912–6926.

Romagnoli, F., Blumeberga, D., & Pilicka, I., (2011). Life cycle assessment of biohydrogen production in photosynthetic processes. *Int. J. Hydrogen Energy, 36*, 7866–7871.

Rozendal, R., Hamelers, H., Euverink, G., Metz, S., & Buisman, C., (2006). Principle and perspectives of hydrogen production through biocatalyzed electrolysis. *Int. J. Hydrogen Energy, 31*, 1632–1640.

Rupprecht, J., Hankamer, B., Mussgnug, J. H., Ananyev, G., Dismukes, C., & Kruse, O., (2006). Perspectives and advances of biological H_2 production in microorganisms. *Appl. Microbiol. Biotechnol., 72*, 442–449.

Sarangi, P. K., & Nanda, S., (2018). Recent developments and challenges of acetone-butanol-ethanol fermentation. In: Sarangi, P. K., Nanda, S., & Mohanty, P., (eds.), *Recent Advancements in Biofuels and Bioenergy Utilization* (pp. 111–123). Springer Nature, Singapore.

Sarangi, P. K., & Nanda, S., (2019a). Bioconversion of agro-wastes into phenolic flavor compounds. In: Sarangi, P. K., & Nanda, S., (eds.), *Biotechnology for Sustainable Energy and Products* (pp. 266–284). IK International Publishing House, India.

Sarangi, P. K., & Nanda, S., (2019b). Valorization of pineapple wastes for biomethane generation. In: Mishra, S., Adhya, T. K., & Ojha, S. K., (eds.), *Biogas Technology* (pp. 169–180). New India Publishing Agency, New Delhi, India.

Sarangi, P. K., & Nanda, S., (2019c). Recent advances in consolidated bioprocessing for microbe-assisted biofuel production. In: Nanda, S., Sarangi, P. K., & Vo, D. V. N., (eds.), *Fuel Processing and Energy Utilization* (pp. 141–157). CRC Press, Boca Raton.

Sarangi, P. K., & Nanda, S., (2020). Biohydrogen production through dark fermentation. *Chem. Eng. Technol., 43*, 601–612.

Sarangi, P. K., (2019). Sustainable utilization of agro wastes using microbial resources: A future perspective. In: Sarangi, P. K., & Sinha, B., (eds.), *Microbe Assisted Sustainable Agriculture* (pp. 257–268). Overseas Press India Private Limited New Delhi, India.

Sarangi, P. K., Nanda, S., & Sahoo, H. P., (2010). Maximization of vanillin production by standardizing different cultural conditions for ferulic acid degradation. *New York Sci. J., 3*, 77–79.

Sarangi, P. K., Nanda, S., & Vo, D. V. N., (2020). Technological advancements in the production and application of biomethanol. In: Nanda, S., Vo, D. V. N., & Sarangi, P. K., (eds.), *Biorefinery of Alternative Resources: Targeting Green Fuels and Platform Chemicals* (pp. 127–139). Springer Nature, Singapore.

Sarangi, P. K., Sahoo, H. P., Ghosh, S., & Mitra, A., *(2009a)*. Isolation of ferulic acid from wheat bran using *Streptomyces* isolates. *Int. J. Appl. Agricul. Res., 4*, 57–62.

Sarangi, P. K., Sahoo, H., Ghosh, S., & Mitra, A., (2009b). Biotransformation of ferulic acid into vanillin by a *Streptomyces* isolate S10. *Asian J. Microbiol., Biotechnol. Environ. Sci., 11*, 273–278.

Sarangi, P. K., Singh, T. A., & Singh, N. J., (2017). Agricultural crop residues: Unused biomass having huge energy potential. In: Ghosal, M., (ed.), *Contemporary Renewable Energy Technologies for Sustainable Agriculture* (pp. 47–61). Narosa Publishing House, New Delhi, India.

Sarangi, P. K., Singh, T. A., & Singh, N. J., (2019). Pineapple as potential crop resource: Perspective and value addition. In: *Food Bioresources and Ethnic Foods of Manipur North-east India* (pp. 83–91). Empyreal Publishing House.

Singh, S., Kumar, R., Setiabudi, H. D., Nanda, S., & Vo, D. V. N., (2018). Advanced synthesis strategies of mesoporous SBA-15 supported catalysts for catalytic reforming applications: A state-of-the-art review. *Appl. Catal A: Gen., 559*, 57–74.

United Nations, (2018). *Sustainable Development Goal 7: Targets*. https://sustainabledevelopment.un.org/sdg7 (accessed on 24 June 2021).

Wakerley, D. W., Kuehnel, M. F., Orchard, K. L., Ly, K. H., Rosser, T. E., & Reisner, E., (2017). Solar-driven reforming of lignocellulose to H_2 with a CdS/CdOx photocatalyst. *Nat. Energy, 2*, 17021.

Wang, A., Sun, D., Cao, G., Wang, H., Ren, N., Wu, W. M., & Logan, B. E., (2011). Integrated hydrogen production process from cellulose by combining dark fermentation, microbial fuel cells, and a microbial electrolysis cell. *Bioresour. Technol., 102*, 4137–4143.

Wang, J., & Wan, W., (2009). Factors influencing fermentative hydrogen production: A review. *Int. J. Hydrogen Energy, 34*, 799–811.

Winkler, M., Hemsehemeier, A., Gotor, C., Melis, A., & Happe, T., (2002). [Fe]-hydrogenase in green algae: Photo-fermentation and hydrogen evolution under sulfur deprivation. *Int J. Hydrogen Energy, 27*, 1431–1439.

Wulf, C., Thormann, L., & Kaltschmitt, M., (2017). Comparative environmental life cycle assessment of biohydrogen production from biomass resources, In: Singh, A., & Rathore, D., (eds.), *Biohydrogen Production: Sustainability of Current Technology and Future Perspective* (pp. 269–289). India, Springer.

Xia, A., Cheng, J., Lin, R., Lu, H., Zhou, J., & Cen, K., (2013). Comparison in dark hydrogen fermentation followed by photo hydrogen fermentation and methanogenesis between protein and carbohydrate compositions in *Nannochloropsis oceanica* biomass. *Bioresour. Technol., 138*, 204–213.

CHAPTER 6

Genetic Engineering and Fabrication of Microbial Cell System for Biohydrogen Production

SUSHMA CHAUHAN,[1] BALASUBRAMANIAN VELRAMAR,[1] RAKESH KUMAR SONI,[2] MOHIT MISHRA,[1] VARGOBI MUKHERJEE,[1] TANUSHREE BALDEO MADAVI,[1] and SUDHEER D.V.N. PAMIDIMARRI[1]

[1]*Amity Institute of Biotechnology, Amity University Chhattisgarh, Raipur, Chhattisgarh, India*
E-mail: spamidimarri@rpr.amity.edu (Sudheer D.V.N. Pamidimarri)

[2]*Department of Biotechnology, Shri Rawatpura Sarkar University, Raipur, Chhattisgarh, India*

ABSTRACT

Hydrogen fuel offers a significant alternative to fossil fuels in the context of a rising carbon footprint in the atmosphere and declining fossil fuel resources. H_2 is the only gas fuel, which is considered clean energy with zero carbon emission upon combustion. Hence, it is encouraged as a replacement of fossil fuels for transportation and industrial applications. However, the process of production is always should be a green process with neutral carbon emission without effluent discharge to extract benefits out of H_2 usage. Thus, the production of hydrogen gas *via* microbial cells is proven a green process without excreting any toxic effluents and shown to be a carbon neutral process. The emission of H_2 gas from the microbial cell is achieved by various mechanisms like photolysis of water by photo-oxygenic method, or hydrogenases generate H_2 by cytoplasmic Hox-hydrogenase enzyme system or released during nitrogen fixation through nitrogenase complexes or *via* dark fermentation through the activity of hydrogenases. In the past, a couple of decades, with the understanding of molecular machinery in microbial cells and technological advancement in molecular biology and microbial system engineering provided tools for researchers to enrich or enhance H_2 production from variable renewable and cheap carbon substrates. The researchers also tried to position the H_2 producing machinery in a non-native H_2 producing host to produce the H_2 by a desirable mechanism. In this book chapter, the efforts were made to discuss the molecular mechanism of H_2 production and enhance its yields by genetic and metabolic engineering of the native host or installing the H_2 machinery in non-H_2 producing host.

6.1 INTRODUCTION

Energy is one of the major components of the national economy, and it is a part of modern human evolution. The rising carbon footprint due to the usage of fossil fuels and their global depletion is forcing us to look for the alternative green fuel sources. In this context, the sources that are considered carbon-neutral or carbon-free are electric power generated through hydraulic turbines, wind, solar energies, and biofuels. Hydrogen fuel upon combustion releases no carbon footprint. However, the thermochemical methods used to synthesize H_2 such as reforming emit greenhouse gases directly or indirectly (Singh et al., 2018). Hence, H_2 cannot be considered a sustainable energy unless it is synthesized using a green process without carbon emissions.

Hydrogen production by biological processes either by microalgae or by dark fermentation is considered a green process since no toxic effluents and carbon footprint are released (Sarangi and Nanda, 2020). Unlike the thermochemical hydrogen production processes, the biological processes are considered to be greener since they are conducted in an aqueous environment at ambient temperature and atmospheric pressure (Kovács et al., 2006). The process of implementing H_2 production *via* the microbial process as a prime source for production is not accelerated. With the advent of molecular biology and the keen interest of the scientific community in biological hydrogen production in the past decade, many technical issues in the area are being addressed (Stephanopoulos, 2007).

To understand the biological hydrogen production, various molecular machinery is characterized by developed models for knowing the key metabolic events in the production of H_2 using the microorganisms. The major H_2 producing machinery is hydrogenases or nitrogenases in the microbial cell. In the microbial system, there are two different types of hydrogen-producing mechanisms such as bio-photolysis and dark fermentation. Both have their advantages and limitations. With the advancement of molecular biology and understanding of the H_2 producing system, many researchers put forward their efforts to address the limitations of different microbial systems for producing H_2. To achieve the enhanced efficiency of H_2 production, the studies are made to fabricate the molecular system in native microbial cells and in non-native hosts by installing the H_2 machinery by using various recombinant microbial system engineering technology. In this chapter, the efforts are put forward to elucidate the mechanism of H_2 production in native microbial cell systems. The discussion is also extended to the tools available for the gene manipulation for microbial engineering and enhancement of target product *via* metabolic engineering or installation of protein machinery.

6.2 HYDROGEN PRODUCTION PROCESSES

Various forms of fossil fuels such as natural gas, petroleum, diesel, and coal are used directly as power and energy sources in the present day. Hydrogen present in the natural gas must be separated from a mixture of the gases already formed or can be produced from natural gases, hydrocarbon, coal, petroleum by various reforming processes. There are few technologies developed to produced H_2 from biomass *via* pyrolysis or gasification (Levin et al., 2010; Okolie et al., 2019; Okolie et al., 2020). Though, waste biomass show less calorific value than coal, the thermal processes favor the production of H_2 (Chattanathan et al., 2012). The merits and demerits of some hydrogen-producing processes are summarized in Table 6.1 (Turner, 2004). Among these processes, the biological process for H_2 production is the most promising because it uses bio-waste as a substrate for H_2 production, and it does not rely on the expensive feedstock or growth medium. Hence, the biological process is gaining more attention in the scientific and industrial sectors (Sinha et al., 2011).

6.2.1 ELECTROCHEMICAL PROCESSES

The electrochemical process or electrolysis is one of the most favorable processes in which electricity is used to fragment the water molecules into hydrogen and oxygen (Stojić et al., 2003). This process takes place in a special chamber also called an electrolyzer. This chamber usually ranges in different sizes, which consists of anode and cathode. In this process, majorly three types of processes are involved such as alkaline electrolyzer, polymer electrolyte membrane electrolyzer and solid oxide electrolyzer. This process has been demonstrated as the potential for renewable energy power generation.

TABLE 6.1 Merits and Demerits of Hydrogen Production by Various Processes

Method	Merits	Demerits
Thermochemical process	• Concentrated H_2 is accumulated • Carbon-rich solid residue is obtained that has many potential applications • The liquid product can be used for biofuel or biochemical recovery • Maximum conversion rate • Feedstock for fuel cell and other hydrocarbon products	• Need electricity input • Cellular impermeability to extract the electrons • CO_2 released into the environment at a high amount. • High effluent discharge
Electrochemical process	• Capable of producing high purity hydrogen. • H_2 is used for fuel cell and O_2 is used for space	• The unit installation cost is very high • Need electricity input
Biological process	• It can produce H_2 directly from water and organic waste • Green algae like cyanobacteria can produce H_2 from H_2O • It can naturally fix N_2 from the environment • Various carbon sources can be used as substrates • This process can use organic waste as a carbon source • It can produce H_2 in the absence of light, e.g., dark fermentation • Dark fermentation is an anaerobic process that does not require O_2	• Some processes required light, e.g., photo-fermentation • O_2 can be harmful to some anaerobic H_2-producing bacteria • Less photochemical efficiency • Nearly 30% O_2 present in gas mixtures • O_2 is a strong inhibitor of hydrogenase • Product gas mixture contains CO_2 which need to be separated • O_2 give inhibitor effect to nitrogenase • Light conversion efficiency is less

This process is considered a green energy source. However, the eco-friendliness of the process mostly depends on the source of electricity used in the process (Marcelo et al., 2008). Besides implementing this method for commercial hydrogen production, the choice of electricity source, efficiency, and cost of the process, as well as emission results from electricity generation, must be considered for economic viability and environmental benefit. Presently, in many countries, electricity is generated *via* the thermal or nuclear process that is not considered as green.

H_2 gas production using electrochemical processes is being continuously used for renewable energy production because this results in zero greenhouse gas emission and reasonably low pollutant discharge. Moreover, the installation of this electrochemical process is highly cost-intensive and is not recommended according to the economic viewpoint for H_2 production. Hence, this process is not considered a process of choice for commercial renewable energy production. Therefore, some recent research aims to overcome the challenges and reduce the cost of the electrolyzer unit. However, there is a long way to overcome these difficulties to make this process a viable, low-cost, and renewable source for H_2.

6.2.2 THERMOCHEMICAL PROCESSES

This process is most of the times used to produce the hydrogen gas from the hydrocarbon source. In this process, the fuel and oxygen are mixed, upon partial oxidation, a hydrogen-rich mixture of gases will be generated, most popularly known as syngas. However, most of the time, this is not considered a green process. There are many strategies designed to make the process, free of environmental hazards and simplify the method to invest less external energy to produce H_2. The physicochemical process relies on the exchange current using catalysts with a graphite-like carbon and an electric field upon forming redox equilibrium between the solutions (Turner, 2004).

The thermochemical process is another most promising process, which uses heat and chemical reaction to release the H_2 from various organic matters such as biomass and fossil fuels. In this process, water molecules can be split into H_2 and O_2 gasses using solar energy or electrolysis. In these methods, the various thermal process uses energy from various resources like natural gas, biomass, and coal to release H_2 (Bonner et al., 1984, Mujeebu, 2016). This process is also well recognized for its efficiency in bulk production with the possibility of large scale-up production. For the thermochemical process, several methodologies have been used to produce H_2, which are mostly natural gas reforming, also called steam methane reforming, thermal dissociation, pyrolysis, gasification, solar thermochemical hydrogen and biomass-derived liquid reforming (Nanda et al., 2017). Among these processes, the thermal dissociation process is one of the well-known processes for its unique mode of action for H_2 production by the direct splitting of water into its corresponding component and produce H_2 (Utgikar et al., 2006).

The thermochemical process is highly suitable for high-temperature solar energy by using concentrated solar power (Genç et al., 2012). Besides, the various combination of methods involved in H_2 production, this process also has demerits with low efficiency for the H_2 production and it requires high-energy consumption. In addition, thermochemical processes require raw material like hydrocarbons and biomass, especially in the case of pyrolysis, gasification or reforming. Hence, for the long-term H_2 production, renewable biomass should be considered.

6.2.3 BIOLOGICAL PROCESSES

Biohydrogen production using various microorganisms, such as bacteria and microalgae, cyanobacteria is a new promising area. This process involves direct and indirect bio-photolysis of H_2O, photo-fermentation, and dark fermentation as well as microbial electrolysis of various organic matters by bacteria. This process is still at an early stage of research. However, in the long term, it could have the potential for sustainable H_2 production with zero or low CO_2 emission.

(i) **Photoproduction of Hydrogen:** The photosynthetic process has been used for the synthesis of carbohydrates as well as biohydrogen by the purple, green sulfur and cyanobacteria. Photosynthesis is the process in which solar light energy is acquired and that is converted into chemical energy in the form of carbohydrates to produce energy. This photoreaction is also capable of releasing hydrogen. The photo bacteria are capable of splitting water molecules to produce electron (e^-) and protons (H^+) in primary photosynthesis. The split e^- and H^+ pass through photosynthetic electron transport chain (ETC) to reach hydrogenase or nitrogenase systems where these enzyme complexes catalyze e^- and H^+ to combine and yield hydrogen gas. In this process, the formation of biological hydrogen depends on the presence of oxygen (Khetkorn et al., 2013). The hydrogenase and nitrogenase enzymes are more sensitive to oxygen (act as a final elector acceptor). The cyanobacteria produce biohydrogen by oxygenic photosynthetic reactions, whereas the purple non-sulfur bacteria accomplish anoxygenic photosynthetic reactions. In anoxygenic photoproduction of hydrogen, the oxygen is not involved as a final electron acceptor.

(ii) **Oxygenic Photoproduction of Hydrogen in Cyanobacteria:** Cyanobacteria are capable of producing the hydrogen through the photosynthetic reactions. Many researchers have well studied the process of photosynthesis in cyanobacteria such as *Synechocystis, Cyanothece, Nostoc,* etc., and elucidated the mechanism of H_2 production during the photosynthesis. The photoproduction of H_2 requires the photons, which are received from the photosynthetic machinery by extracting solar energy. All these reactions are conducted in vegetative cells. The photosynthetic machinery consists of a light-harvesting system and electron transporting complexes (Jordan et al., 2001; Zouni et al., 2001). Phycoerythrin (PE) and phycocyanin (PC) absorb rays near 490 nm and 580 nm, respectively, and as combined they are called phycobilisome (Grossman et al., 1993). Phycobilisomes act as a light-harvesting system for cyanobacteria. These characters indicate that cyanobacteria selectively colonize in the illuminated habitat.

The photon energy absorbed from the sunlight by the photosystem-II (PS-II) and then transferred to the photosystem-I (PS-I) complex (Hall and Rao, 1999; Blankenship, 2002) through the redox reactions. The excited energy is converted and transferred to the water molecule to generate molecular O_2 and e^- by O_2 evolving complex (OEC) of Mn_4O_4Ca-cluster in the PS-II complex (Kern et al., 2007, Barber 2008). The electrons are subsequently shifted through the plastoquinone (PQ) complex, cytochrome-b6f complex and plastocyanin. PS-I creates the electron movement that reduces the iron-sulfur protein ferredoxin (Fd). Furthermore, the Fd passes the electron to $NADP^+$ and form NADPH by the ferredoxin-NADP-reductase enzyme. The molecular O_2 and protons are released into the lumen of the chloroplast and make H^+ gradient in intermembrane space. This gradient force will help to produce high-energy molecule ATP through ATP synthase and generate reduced NADPH and is necessary to synthesis the carbohydrates by using CO_2. This reduction of Fd and NADPH is recycled for the production of hydrogen by the hydrogenase enzyme. The hydrogenase enzyme stimulates the reverse reaction of joining the e^- and H^+ to synthesis H_2.

This production method will not be a promising procedure in large-scale biological synthesis of H_2. Even though the dark fermentation of *Cyanothece* has been grown with minimal media without nitrogen source enhanced the activity of the hydrogenase enzyme under a low oxygen environment, the net production of H_2 increased two-fold in this condition. This final productivity was found to be insignificant from an economic perspective (Skizim et al., 2012).

Another specific reaction called nitrogen fixation is a characteristic reaction that takes place in cyanobacteria in a specialized cell named heterocyst. These heterocysts are involved in the nitrogen fixation by the oxygen-sensitive enzyme nitrogenase. Nitrogenase is a multipart protein catalyzes the conversion of atmospheric molecular N_2 to NH_3 (Igarashi et al., 2003). The endergonic reaction requires ATP and the molecular reaction is as follow:

$$N_2 + 8H^+ + 8e^- + 16ATP \rightarrow 2NH_3 + 16ADP + 16Pi + H_2\uparrow \quad (6.1)$$

During the fixation of dinitrogen, the hydrogen is generated as an additional product. The N_2-free environment also expands the hydrogen production significantly by following chemical reaction and this is catalyzed by a bidirectional Hox-hydrogenase. However, this enzyme complex catalyzes both synthesis and uptake reactions.

$$2H^+ + 2e^- + 4ATP \rightarrow H_2\uparrow + 4ADP + 4Pi \quad (6.2)$$

The Hup-hydrogenase is an enzyme present in the thylakoid membrane in almost all the cyanobacteria in the heterocysts and consumes the hydrogen released by nitrogenase. Hence, in many cases, the net gain of H_2 release from the cell is very low. This is an aspect to be considered for the engineering cell with no or lowHup-hydrogenase activity. Hup-hydrogenase consists of two subunits HupS (small subunit) and HupL (large subunit). HupL consists of a catalytic unit; hence, many researchers targeted this subunit either for blocking *via* chemical means or physical knockouts (Khetkorn et al., 2013). Hox-hydrogenase, which catalyzes the bidirectional reactions of both H_2 synthesis and consumption, is a pentameric NAD^+ reducing enzyme, which is coded by *hoxEFUYH*. This enzyme complex includes two protein complexes; one is hydrogenase complex encoded by genes *hoxY* and *hoxH* and diaphorase complex coded by *hoxE, hoxF,* and *hoxU*. The significant physiological role of this bidirectional enzyme in the cell is not very clear. However, it has a role as an electron valve during photosynthesis. Any defects in this gene leads to oxidation of PS-I and causes higher fluorescence of PS-II. It is also known for the mediator in buffering the excess of reducing power generated during anaerobic conditions and maintains active fermentation conditions (Cournac et al., 2004).

In contrast, the development of vegetative cells to heterocyst is prohibited in the N_2 free condition by *hetR* gene produces HetR, a serine-type protease by which the heterocyst differentiation is determined at an early stage (Khetkorn et al., 2013). At the same time, the heterocyst formation is remarkably high leads to nitrogen fixation increase many folds. By metabolic engineering, increasing the copies of *the hetR* gene as well as knocking out of *hup* gene will improve the hydrogen production in cyanobacteria, and the resulting strain has good industrial applications. The detailed mechanism of the H_2 production is elucidated in Figure 6.1.

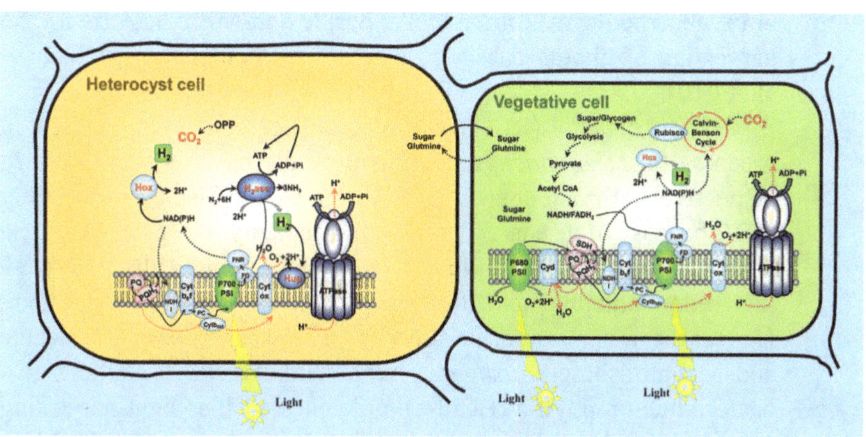

FIGURE 6.1 Molecular pathway cyanobacteria follow for the production of H_2 via oxygenic and non-oxygenic conditions by vegetative and heterocyst cells, respectively. (Modified from Khetkorn et al., 2013).

The metabolic pathway in the vegetative cells follows the oxygenic pathway and heterocystous cyanobacteria contain heterocyst differentiation for nitrogen-fixing to have a non-oxygenic pathway. Cyd, Quinol oxidase; Cyt b6f, Cytochrome b6f; Cyt c553, Cytochrome c553; Cyt ox, Cytochrome c oxidase; Fd, Ferredoxin; FNR,

Ferredoxin-NADP reductase; Hox, BidirectionalHox-hydrogenase; Hup, Uptake hydrogenase; N2ase, Nitrogenase; NDH-I, NADPH dehydrogenase (complex I); OPP, Oxidative pentose phosphate pathway; PC, Plastocyanin; PS, Photosystem; PQ, Plastoquinone pool; Rubisco, Ribulose-1,5-bisphosphate carboxylase oxygenase; SDH, Succinate dehydrogenase. Proton movement is given in red dotted arrow, black arrows represent the electron flow.

(iii) Purple Bacterial Photoproduction of Hydrogen: Purple photosynthetic bacteria, which are distinctively anoxygenic photosynthetic bacteria, are of two types such as purple sulfur and purple non-sulfur bacteria. Purple sulfur bacteria use sulfide and hydrogen as an electron-donating molecule, whereas purple non-sulfur bacteria use different organic molecules (Madigan et al., 2008). The purple bacteria synthesize their energy molecules from solar energy by photosynthesis. This photosynthesis process takes place in the absence of oxygen, termed as anoxygenic photosynthesis. In contrast to cyanobacteria, the near-infrared spectral range between 800–870 nm is absorbed in maximum by the purple non-sulfur bacteria for the harvesting of photon energy to produce molecular energy (Saga et al., 2010).

Reaction center-1 (RC-1) and Reaction center-2 (RC-2) present in green sulfur (*Heliobacteriaceae*) and purple bacteria (*Chloroflexi*), respectively (Bryant et al., 2006). In this chapter, we discussed only RC-2 because of its energy transferring capacity. The purple bacteria have 870 nm (P870) absorbing photosystem alone with moderately low redox potential and absence of water splitting photosystem. Hence, the water cannot be used as an electron-donating molecule and accomplishing anoxygenic photosynthesis. P870 contains two bacterial chlorophyll-a (BChl-a) molecules such as light-harvesting complex-1 and 2 (LH-1 and 2). The photons received by the LH-1 and 2 and excited electrons transfer to bacteriopheophytin. The e^- and protons reduce the menaquinone and form Menaquinol (QH_2). Then, the electrons are subsequently pass through the cytochrome bc1 complex. Finally, the electrons return to RC-2 through cytochrome called cyclic electron transport to create a proton gradient for the production of ATP synthesis. The blocking of this cyclic electron transfer creates a reducing equivalent. Hence, the BChl-a runs in a positively charged condition, should be reduced to uptake the e^- from the other sources to fulfill the e^- requirement at $P870^+$. That can be achieved by the utilization of organic molecules. The purple non-sulfur bacteria can employ the N_2 gas to produce the NH_3 for their survival and reproduction.

The nitrogenase and hydrogenase enzymes can be explored to produce biological H_2. The advantage of purple non-sulfur bacteria is performing photosynthesis in the absence of O_2. At the same time, the presence of O_2 leads to inhibit the normal function of nitrogenase and hydrogenase enzymes. These microorganisms are used for the biological production of H_2. The biotechnological approach could give promising results by blocking the expression of important genes and their functions as follows, suppression of the uptake hydrogenase, increase in nitrogenase enzyme production, blocking of competitive pathways like PHA production and accumulation inside the cells, reduce the pigment production especially BChl-a.

(iv) **Non-Oxygenic Hydrogen Production via Dark Fermentation:** Non-oxygenic H_2 production does not require any light energy as the fermentation operates in dark condition. This makes the system easier and needs less investment in designing the reactor system. Hence, it is more suitable for commercial production. The two groups of anaerobes, the obligate and facultative microorganisms grow in the absence of O_2 are capable of producing H_2. In the case of strict anaerobes, the electrons are harvested through pyruvate oxidation and then these electrons are utilized for the oxidation of ferredoxin (Fd). Further, these electrons move through the hydrogenase and are harvested by protons to form H_2. The prominent species like *Clostridium, Ethanoligenens,* and *Desulfovibrio* produce hydrogen by this pathway. The enteric bacteria like *Enterobacter, Citrobacter, Klebsiella, Escherichia coli* and *Bacillus* spp. produce H_2 via the formate hydrogen lyase system (Brosseau et al., 1982). Detailed pathways and patterns of metabolism for hydrogen production are presented in Figure 6.2. In the dark fermentation, only a pair of electrons will sink for H_2 synthesis. The remaining electrons end up in producing organic acids and other products. Hence, from a physiological point of view, H_2 is a by-product of dark fermentation rather than the main product. Hence, much of understanding is needed for the engineering the cellular flux towards H_2 production.

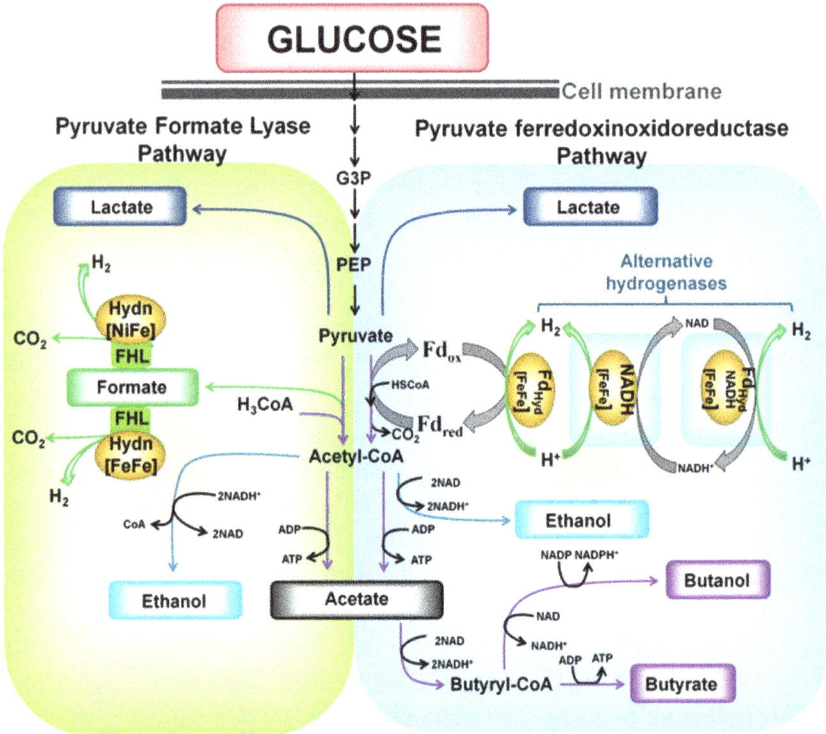

FIGURE 6.2 H_2 production pathways in dark fermentation followed by different bacterial species using glucose as basic carbon sugar.

Taking the glucose as the basic carbon sugar and metabolic starter for the cell, the majority of the microorganisms proceed to the basic pathway, i.e., glycolysis for the production of pyruvate. This pyruvate is converted to either acetyl-CoA and CO_2 or either acetyl-CoA and formate. In the former situation, the cell will reduce the ferredoxin and then upon the action of pyruvate ferredoxin oxidoreductase (PFOR) will produce H_2. In the latter case, with the action of formate hydrogen lyase (FHL) protein complex formate is converted into H_2 and CO_2. This overall pathway is popularly known as the pyruvate formate lyase (PFL) pathway. The most popular organism, which follows the PFOR, is *Clostridium* sp., and *E. coli* follows the latter. These bacteria are considered model organisms to study and understand the two major pathways. The molar productivity of these pathways differs significantly. The FHL system use [FeFe]-hydrogenase and does not use the NADH produced during glycolysis and accumulate ethanol or lactate as the end product. Hence, the final productivity of this pathway is 2 mol of H_2 for 1 mol of glucose. Unlike the FHL system, in the PFOR pathway, the hydrogen is released by the oxidation of reduced ferredoxin (Fd_{red}) with the help of ferredoxin dependent [FeFe]-hydrogenase. An advantage of this system is that 2 more molecules of H_2 can be generated via oxidation of NADH with the help of NAD-dependent [FeFe]-hydrogenase. Hence, the final productivity could reach up to 2–4 moles of H_2 from 1 mol of glucose (Tapia-Venegas et al., 2015).

6.3 ENGINEERED MICROBIAL SYSTEMS FOR HYDROGEN PRODUCTION

In the past couple of decades, with the advancement in the genomics studies and the development of various molecular tools for the genetic engineering of microorganisms has helped the researchers to engineer the microbial system to enhance H_2 productivity. The advanced tools like transcriptome analysis, next-generation sequencing, RNA sequencing technology, ligation free cloning, golden gateway cloning, Gibson assembly, various advanced polymerase chain reaction (PCR) and protein analysis tools can facilitate high throughput engineering of native strains to enhance H_2 production. The enhancement of productivity in native H_2 producing hosts can be obtained. The fabrication of the non-H_2 producing host by installing the molecular machinery into the cell can become possible to generate H_2.

As described in the earlier sections, there are two major mechanisms available for H_2 production (i.e., photolysis of water and dark fermentation) by cyanobacteria and eubacteria, respectively. Many researchers have explored both the systems for molecular engineering to enhance H_2 productivity. In this section, a detailed discussion will be made on the efforts made on genetic engineering and/or fabrication of microbial cell, which includes both cyanobacterial and eubacterial systems (Oh et al., 2013, Knoot et al., 2018).

6.3.1 ENGINEERING OF CYANOBACTERIAL CELLULAR SYSTEM FOR HYDROGEN PRODUCTION

Cyanobacteria are microbial species that perform photobiological hydrogen evolution via two different methods. One is by creating H_2 as a by-product during nitrogen fixation by nitrogenases, the other is by anaerobic dark fermentation. Nitrogenase catalyzes the fixation of nitrogen by converting it into ammonia. For the fabrication of cyanobacterial cell systems for better performance in hydrogen production or strain improvement, the majority of studies were made in species like *Synechocystis*, *Anabaena variabilis*, *Anabaena* PCC7120 and *Nostoc punctiforme*. The reason for this selection is the availability of the following (Heidorn et al., 2011, Huang et al., 2013):

- Annotated genome sequence information available in the public database.
- The cellular mechanism is well studied and molecular tools and protocols are relatively available compared to other cyanobacterial species.
- Their growth conditions are ambient and easily cultivable in the lab environment.
- Easy for the genetic manipulations since vectors and advanced genetic engineering tools are already available for these species.
- Scale-up from lab scale to industrial scale is easy since they require minimal nutrients for their metabolism and growth.

The encouraging results in hydrogen production from cyanobacteria, a new model organism has come into light, i.e., *Cyanothaceace* ATCC 51142—a strain with high efficiency of H_2 production *via* photosynthetic photolysis of water (Bandyopadhyay et al., 2010). It is presently considered as a model system for the research in the H_2 production, and many genetic tools were developed to understand and to further enhance the H_2 production. This work has paved the way for developing cyanobacterial synthetic biology and biosystem engineering. In this regard, many vectors, promoters, ribosomal binding sites, terminators, transcriptional, and translational controlling units were come into the picture and assisted in the understanding of the cyanobacterial cell and developing them as cell factories for fuel and fine chemicals for various applications (Khan et al., 2019). Two major areas have been focused in the case of cyanobacteria for the enhancement of H_2 production such as:

- Metabolic engineering of target strains for diverting the metabolic flux towards enhancing H_2 production
- Engineering enzyme complexes like hydrogenases and nitrogenases to enhance the hydrogen production.

(i) Hydrogen production enhancement via diverting the electron flow system using chemical inhibitors

In cellular respiration, the electrons donated by electron donors (e.g., NADPH) are passed across the membrane by electron transport chain will be accepted and transferred to the membrane. Diverting the electron passing through the electron transport chain towards the hydrogenase or nitrogenase enzyme complex will be accepted by proton to form H_2. Hence, a decrease of electron flow through the electron transport chain across the membrane leads to an increase in H_2 evolution by the cells. Skizim et al. (2011) explored this strategy. The authors used chemical inhibitors for blocking the NADH dehydrogenase complex type II resulted in a two-fold increase in hydrogen production. However, the process was conducted in dark conditions. In another case with *A. Siamensis*, using the rotenone as an inhibitor for NADH dehydrogenase under light conditions could increase Hox-hydrogenase activity under light conditions resulted in enhancement of the H_2 accumulation. In the same way, using malonate, the inhibitor of succinate dehydrogenase, resulted in the enhancement of hydrogen production dark conditions in *Synechocystis* PCC 6803 (Burrows et al., 2011). In contrast, malonate could not show any betterment in hydrogen production compared to control in the presence of light. This is because succinate dehydrogenase is not a pool of electron sources. The inhibition of terminal oxidases participate in the respiratory event can yield electrons and it could participate in the H_2 production.

Khetkorn et al. (2012) in their study used various specific inhibitors like potassium cyanide, rotenone, DCMU (3-(3,4-dichlorophenyl)-1,1-dimethylurea) and DL-glyceraldehyde. This study is subjected to redirect the electron flow towards the nitrogenase and biodirectional Hox-hydrogenase. The attempt successfully enhanced the H_2 production by the cyanobacterium *Anabaena siamensis* TISTR 8012 (Khetkorn et al., 2012).

The electrons can also be diverted towards nitrogenase and Hox-hydrogenase from the photosynthetic system. The oxygenic photosynthesis generates molecular oxygen *via* photolysis of water, which can inhibit the nitrogenase activity and further hydrogen production. Hence, the application of inhibitors of water splitting photosystem II will help in enhancing the evolution of H_2 from the cell. With this strategy, Weissman, and Benemann (1977) studied the application of DCMU in H_2 production by enhancing the nitrogenase activity. Although using specific inhibitors enhanced hydrogen production, the concentrations were finely adjusted. The abnormal concentrations will lead to an adverse effect on growth and productivity.

Moreover, this enhancement via chemical additives is not a suitable method for commercialization.

(ii) Enhancing hydrogen production by inhibition of cellular uptake

In the context of hydrogen production enhancement, the primary criteria are the mechanism of hydrogenases to produce H_2. The native hydrogenases have the bidirectional catalytic ability where hydrogenases produce H_2 and consume it back for the reducing reactions in the cell. Hence, the primary focus was made to disrupt the H_2 uptake by hydrogenases so that the cells can accumulate more H_2. This can be achieved by chemical blocking of the enzyme complex to hinder the H_2 uptake or genetically modify the enzyme complex to stop the uptake of H_2 accumulated in the medium. However, the prior strategy of blocking the H_2 utilization by cell using chemicals is not advised for industrial-scale since it needs to be maintained in molar concentration to achieve the accumulation without utilizing back by the cells. The later one looks promising since it is by a targeted genetic change where the uptake reaction is blocked *via* knockout or site-directed mutagenesis of the catalytic region of the enzyme.

As stated earlier, in cyanobacterial species, there are two distinct types of hydrogenases are known as Hup and Hox. Among these, Hoxcatalyses both uptake and production of H_2, whereas the Hupcatalyses unidirectional uptake. Masukawa et al. (2002) studied the effect of these genes upon inactivation in *Anabaena* sp. PCC 7120. In their study, the single and double mutants of Hox and Hup are generated, and upon removing, the activity of Hup resulted in a seven-fold increase in the rates of H_2 release from the mutant cells compared to wild type. This study resulted in the application of this strategy for enhancing H_2 production in multiple other species like *Anabaena* and *Nostoc* strains (Schütz et al., 2004; Yoshino et al., 2007).

There are multiple studies conducted where deficient mutants of various genes were tried in cyanobacterial species for H_2 enhancement and most significant among is deficient mutants achieved *via* advanced gene knockout strategies. Among these deficient mutants, the mutants of *A. variabilis* AVM13 (Happe et al., 2000), *N. punctiforme* NHM5 (Lindberg et al., 2002), *Anabaena* PCC 7120, *Anabaena* PCC 7120 (Zhang et al., 2014), *Nostoc*PCC 7422 (Lindberg et al., 2012), and *A. siamensis* (Khetkorn et al., 2012) resulted in enhanced accumulation of the H_2 compared with wild-types. All these studies were successful in inhibiting the hydrogen uptake by cells, and most of the studies observed that no effect in nitrogen-fixing ability. Among the above studies, a significant observation made by Khetkomet al. (2012) was that by inactivation, the gene *hupS* by knockout process achieved approximately four-fold enhancement in H_2 production and at the same time nitrogen fixation as much as 12-fold more than the wild-type strain.

(iii) Enhancing hydrogen production by cell-system engineering

There are multiple ways possible for enhancing H_2 production. The first aspect that could be considered is increasing the substrate availability and the second could be diverting electron flow towards hydrogen production. In the cyanobacterial system, the substrate for the H_2 is the availability of reduced NADPH. In the cellular system, the electrons are transferred from the photosynthetic system to other assimilatory pathways make them inadequately available for nitrogenases and Hox-hydrogenases for H_2 production. Hence, in recent years the efforts were put forward in removing the competing electron pathway, and divert them towards H_2 production machinery.

Electron donors like NADPH and ferredoxin act as substrates for the proton generation and need electrons for the generation of H_2 by nitrogenase or Hox-hydrogenase. However, NADPH is also a substrate for the respiratory complex I coded by gene *nad1* which codes for NADPH-dehydrogenase (NAD1). Eliminating the NAD1 can make an electron available for the H_2 production. The above-stated strategy was successfully studied in the *Synechocystis* M55. In this study, the activity of NAD1 was eliminated by knocking out the gene *ndhB,* which codes for the large subunit of the NAD1 protein (NdhB) resulted in enhancing the H_2 production up to 50%. Moreover, uptake activity was also reduced to a good extent (Cooley et al., 2001). The research studies are also put forward to see the effect of removing all respiratory terminal oxidases (Δ*ctaI*, Δ*ctaII* and Δ*cyd*) from the cell. The recombinant strain able to enhance H_2 production. The same strain when quinol oxidase gene activity was eliminated could enhance the bidirectional Hox-hydrogenase activity to several folds in the presence of light. The other electron assimilatory pathway, which reduces the availability of electrons for H_2 production, is the nitrate assimilation pathway. Hence, the efforts were made to generate the recombinant strain with removed nitrate assimilation pathway were tried in multiple studies. The significant among those is removing the activity of nitrate reductase (Δ*narB*) or nitrite reductase (Δ*nirA*) or both the genes (Δ*narB*/Δ*nirA*) in *Synechocytis* PCC 6803. The resulted strain could able to enhance H_2 production compared to wild-type (Baebprasert et al., 2011).

The majority of H_2 production depends on enzyme complexes, which conduct the catalysis of proton combining with electrons to release H_2. As we discussed this part in earlier sections, it is clear that nitrogenases or Hox-hydrogenases are responsible for the H_2 production. Since the cyanobacteria conduct the oxygenic photosynthesis process for the production of biomass and energy harvesting, the released oxygen inhibits the H_2 production. However, many cyanobacterial species can perform anaerobic dark fermentation in the absence of light using the carbohydrates synthesized and stored during the light period. This process can also be used for the H_2 production in the absence of light. Many cyanobacteria metabolize the carbohydrate substrates to perform the anaerobic fermentation to produce H_2 and accumulate the lactate, ethanol, acetate, and CO_2 as by-products. To enhance better H_2 productivity in the dark fermentation, one can eliminate the key gene/genes to move the metabolic flux towards hydrogen production. The best example of this model is, by eliminating *IdhA* (*ΔldhA*) codes for the lactate dehydrogenase, this enzyme is the NADPH dependent and converts pyruvate to D-lactate. The *Synechococcus* PCC 7002 with a lack of *IdhA* coding gene, showed a higher ratio of NADPH to $NADP^+$ in the cell and could be able to release fivefold more H_2 than the wild type.

(iv) Molecular engineering of hydrogenases and nitrogenases

In cyanobacteria, the main machinery involves in H_2 generation is hydrogenases and nitrogenases. It is well known that the rate of the production of H_2 is very less. Hence, the researchers looked for the strategy of engineering hydrogenases and nitrogenases in the native host system or transferring the efficient hydrogenases to a heterologous system for improving productivity. Nitrogenase is one of the components of H_2 production. However, the low turnover productivity (1 mol of H_2 per mol of nitrogenase) can use two molecules of cellular ATP for every electron reduction is an energy-intensive process for the cell. In another way, the hydrogenases have better productivity like in the case of [FeFe]-hydrogenase can evolve a high rate of hydrogen with no input of cellular ATP. Moreover, the conserved structure and simple maturation pathway, [FeFe]-hydrogenase is a good choice for transfer to a heterologous system. This was tried expressing the [FeFe]-hydrogenase from dark fermentative bacteria *Clostridium* in *Synechococcus* PCC7942 (Asada et al., 2000). However, the hydrogenase did not effectively express. The authors discussed the importance of protein maturation in the active expression of hydrogenases for the effective installation of protein machinery in the cyanobacterial cell (Asada et al., 2000).

In a separate study, Ducat et al. (2011) studied the expression of hydrogenase HydA from *Clostridium acetobutylicum* in the non-nitrogen-fixing cyanobacterium *Synechococcus elongates* sp. 7942. In this study, the authors demonstrated successful heterologous expression of active hydrogenase in a non-native host and are found to act as light-dependent hydrogenase could produce hydrogen under anoxic conditions (Ducat et al., 2011). H_2 productivity is almost 500-fold greater than control. The authors also concluded that adding the external ferredoxins to the media could modulate the redox flux in the recombinant strain expressing the *Clostridium* hydrogenase, which could be able to yield higher hydrogen.

The successful heterologous expression of hydrogenases in the cyanobacterial system still remains a challenge because of the limited information available related to the structural and maturation of the active hydrogenase and their installation in the native host. In a very recent study instead of heterologously expressing the hydrogenase from another organism, Gutekunst et al. (2018) studied the effect of enhanced turnover number in the cell of native host cells suggested the molar concentration of hydrogenase in the cell is proportional to the amount of hydrogen released. In this work, the authors were successfully able to over-express the hydrogenases to increase the hydrogenase molecules up to 400–2000 in a cell. The *in-vivo* photosynthetic electron transfer rates are found to be 47 e^-/s and productivity equivalent to 25 H_2/s. The strategies used by the authors in the study could help to design novel overexpression strategies for H_2 production.

Oxygen sensitivity is one of the important hindering factors for H_2 production *via* photocatalytic way. These hydrogenases in cyanobacteria either Hox-hydrogenase or nitrogenase both show sensitivity toward oxygen. At least, the tolerance to some extent by oxygen released during the photosynthetic activity should be considered for making a light-driven H_2 production that would give origin to an efficient photo-biological-based microbial fuel cell that generates H_2. Hence, the oxygen sensitivity of hydrogenases should be bypassed. Thus, many researchers searched naturally oxygen tolerant hydrogenases for understanding and use knowledge for engineering the oxygen-sensitive hydrogenases to tolerant and/or resistant *via* recombinant methods.

In the literature, several microorganisms are reported to have [NiFe]-hydrogenases showed oxygen tolerance in using H_2 as the sole source of energy for reducing the $NADP^+$. In addition, the studies showed that [NiFe]-hydrogenases show oxygen tolerance to a certain extent, whereas [FeFe]-hydrogenases were found to be oxygen intolerant. The best example in this category is [NiFe]-hyrogenase of *Ralstoniaeutropha*. The enzyme complex is considered a model system for studying oxygen tolerance in hydrogenases. The studies demonstrated that the tolerance of the hydrogenase towards the oxygen is achieved by structural conformational change at FeS clusters. Realizing the importance of [NiFe]-hydrogenase, Weyman and his co-workers overexpressed the bidirectional oxygen tolerant [NiFe]-hydrogenase from *Thiocapsaroseopersicina* along with accessory proteins in the cyanobacterium *Synechococcus* PCC 7942 (Weyman et al., 2011). The authors concluded that the active expression in the hydrogenase in the host cell and H_2 release *in-vitro* studies. However, no H_2 released from the cell *in-vivo* is reported. However, practically for the application in industrial scale, this strategy could not be economically feasible. Moreover, this will add the operation cost and further increase the production cost. Hence, the engineering of nitrogenase is a practical alternative to divert the electron towards H_2 production.

Masukawa et al. (2007) generated a series of mutants, which cannot fix the nitrogen by *Anabaena* sp. 7120. The recombinant cells carrying mutation *ΔHup, ΔNifV1* resulted in a higher rate of H_2 production in the presence of N_2. Though the gene accessory proteins for the nitrogen fixation can diminish the nitrogenase activity, the substitution of amino acids in the active center can lead to the loss of activity without disturbing the hydrogen production by nitrogenase. Amino acid replacement of *Azotobacter vinelandii* nitrogenase catalytic center eliminated N_2 fixation and allowed better proton reduction, helped in more H_2 production in the presence of N_2, and the aerobic environment in the case of cyanobacteria. Likewise, several variants generated by amino acid replacements in the catalytic site of nitrogenase in *Anabaena* exhibited the increased levels of H_2 production compared to the reference strains in the presence of N_2 (Masukawa et al., 2010).

Cyanobacteria appear to be model microbial factories for the H_2 production *via* the photo-biological catalytic system. However, to design a perfect photosynthetic microbial fuel cell for the H_2 production is still a long way to run. The prerequisite of achieving this objective, more studies are needed for understanding the natural system for H_2 production. A suitable model organism needs to be developed for the studies focusing on cyanobacteria looking forward to engineering and develop photosynthetic microalgae-based microbial fuel cells. This will enhance the cyanobacteria producing H_2, the green and renewable way adding no carbon footprint to the environment.

6.3.2 GENETIC ENGINEERING OF DARK FERMENTING BACTERIA FOR HYDROGEN PRODUCTION

The major importance of dark fermentation is the ease of cultivation and controllable condition for growth. Hence, dark fermentation is acquiring significance as the most probable biological method considered for H_2 production. The significance of this system is, simple sugars and organic derivatives help as carbon source. Hence, the system is easily adaptable for commercial scale-up. With the developments in the molecular biology techniques, the availability of genome sequences in public databank facilitated an understanding of molecular mechanisms from different bacterial dark fermentation machinery for the production of H_2. Hence, many researchers put forwards their efforts to engineer various cellular machinery and metabolism to make the cellular metabolic flux divert towards enhancing H_2 production in the native bacterial cell. The studies were also made to install the H_2 producing machinery in non-native hydrogen-producing bacteria, which has other fermentative advantages. In this section, we will discuss in detail about the studies until now carried on bacterial cell system engineering for H_2 production.

Different bacteria produce H_2 by two major methods. The two enzyme complexes involved in the process are pyruvate:ferredoxin oxidoreductases [PFOR] and pyruvate:formate lyase [PFL]. The basic reaction involved in the biocatalysis of these two enzyme complexes is shown in Eq. (6.3) and Eq. (6.4). In the PFL system produce hydrogen with the complex hydrogen lyase called formate:hydrogen lyase.

$$\text{Pyruvate} + \text{CoA} \rightarrow \text{Acetyl-CoA} + \text{Formate [PFL]} \tag{6.3}$$

$$\text{Formate} \rightarrow H_2 + CO_2 \text{ [FHL]} \tag{6.4}$$

In the system, pyruvate:ferredoxin oxidoreductases (PFOR) combined with Fd-dependent hydrogenase (HydA) produce the hydrogen in the cell, and the reaction is carried by these complexes are presented in the Eq. (6.5) and Eq. (6.6) (Hallenbeck et al., 2009).

$$\text{Pyruvate} + \text{CoA} + 2\ \text{Fd}_{ox} \leftrightarrow \text{Acetyl-CoA} + CO_2 + 2\text{Fd}_{rd} \quad \text{[PFOR]} \tag{6.5}$$

$$2\text{Fd}_{rd} + 2\ \text{H+} \rightarrow 2\ \text{Fd}_{ox} + H_2 \text{ [HydA]} \tag{6.6}$$

Under the dark fermentation, though the theoretical yields of H_2 are up to 12 moles of H_2 per mole of glucose. However, the reported highest yields achieved in the native host is less than 4 mol/mol of glucose. Hence, there could be a possible window for improving the H_2 productivity in the native host through genetic engineering using modern biotechnology tools. To make the biological hydrogen production economic and competitive, there are two major strategies could be implied. One is, engineering the native strain to use the sugars and other organic molecules, which are low cost or derived from waste biomass (e.g., lignocellulosic biomass, kitchen waste, industrial waste, etc.). The later could be engineering the H_2 producing machinery to reach near to the theoretical stoichiometry. Both choices could be helpful in strain engineering for making the process economic and feasible at an industrial scale.

(i) Metabolic engineering of native bacterial host for hydrogen enhancement

Extending the metabolism towards low-cost carbon source, there are many aspects of metabolic engineering were studied to enhance the H_2 production. Under this section, we are going to discuss extending native bacterial hosts to incorporate the assimilation ability of different sugars into cellular metabolism. Bacteria are found to use a wide range of carbon in the form of sugars and incorporate them into their metabolism. However, every limited number of bacterial strains are reported to use carbon sources other than glucose for H_2 production. It can be cost-effective if the cell uses the sugars derived from the waste biomass like lignocellulosic biomass.

There are bacterial species like *Clostridium* sp., other thermophilic bacteria that could directly use the complex carbon sources like cellulose, hemicellulose, and starch as either sole carbon or mixed form, and can produce H_2. However, these are not directly useful for commercial applications since they need thermophilic conditions for growth, and their growth rate is very low, and the production rate is very limited in these substrates. Hence, the production cost of the unit of H_2 will be very high and could not be implemented in the commercial sector unless the efficiency of H_2 production is increased to several folds to reach possible near to theoretical values. Hence, the major focus is put forward to integrate the enzyme related to the metabolism of complex sugars and removing or blocking several genes to divert metabolic flux towards H_2 production with the application of recombinant DNA technology.

Clostridium sp., a well-known obligate anaerobe, uses PFOR for H_2 production, which has the production ability of theoretical maximal 4 mol/mol of glucose. However, the experimental value of H_2 production is far less than theoretical values; hence, this area is identified as a possible window for improvement. Thus, many researchers focused on metabolic engineering or installing the H_2 producing machinery towards reaching better efficiency. Marimoto et al. (2005) in their study improved the H_2 production to 2.4 mol/

mol glucose, whereas the wild type could produce less than 1.4 mol/mol glucose. This was achieved by overexpressing the [FeFe]-hydrogenase in *C. Paraputrificum* M-21. The same strategy was implied in the strain *C. tyrobutyricum* JM1 where HydA was overexpressed and recombinant could able to enhance yields up to 1.5 fold in glucose as a carbon source compared to wild type.

In another study, Lie et al. (2006) improved the H_2 production in *C. tyrobutyricum* by deleting the gene *ack*, which codes for the enzyme acetate kinase. The productivity was improved to 1.5 fold in the recombinant knockout strain compared to the wild type. These studies not only helped in improving the H_2 production but also understanding the H_2 production mechanism. With the help of this knowledge, there are a series of knockouts developed in this regard named ClosTron strains for *Clostridium* sp. such as *C. botulinum, C. acetobutylicum, C. sporogenes* and *C. difficile*. These strains were developed by Heap et al. (2007, 2010) improved our understanding of carbon metabolism of *Clostridium* sp. and H_2 production pathways.

Many enteric bacterial species can produce H_2 *via* PFL/FHL system. Based on various studies, it is also proposed that the species like *Enterobacter aerogenes* have an additional system that is based on the putative membrane-bound hydrogenase depend on the NADH-dependent pathway (Nakashimada et al., 2002; Zhang et al., 2009; Lu et al., 2010). To the proof of concept Nakashimada et al. (2002) showed NADP-dependent release of H_2 by demonstrating the *in-vitro* release of membrane fraction isolated from *E. aerogenes* AY-2.

Zhao et al. (2009) in his study he developed the knockout recombinant of *E. aerogenes* IAM1183 strain with lack of gene activity of *hycA*, which codes for the FHL repressor. The results found that the yield was significantly improved compared to the wild type (Zhao et al., 2009). Zhang et al. (2009) studies showed that the simple addition of NADH or NAD^+ into the medium can increase the H_2 yields due to change in the intracellular redox state in the *E. aerogenes* (Zhang et al., 2009). In continuation of this concept, Lu et al. (2010) has developed the knockout for NADH-using lactate dehydrogenase and overexpression of NAD^+-dependent formate dehydrogenase and successfully improved the H_2 production. However, much of the cellular coordination events, which could improve the H_2 production need to be understood to enhance the H_2 production at least near to possible theoretical values.

(ii) Engineering of *E. coli* for hydrogen production

The majority of understanding and the knowledge of microbial cells are originated from studies in *E. coli*. The inherent growth characteristics and ease of cell manipulation, *E. coli* became the model system to study many molecular aspects of the cell from late 1960. With the advancement in molecular techniques and various tools developed for the easy molecular fabrication of cells and established protocols for genome engineering making *E. coli* a suitable model for the various cellular process.

E. coli being an enteric bacterium has the inbuilt cellular machinery for hydrogen production. Hence, *E. coli* also helped in understanding the cellular process for H_2 production from the ground level. These studies not only helped in gaining the basic knowledge on biochemical and thermodynamic aspects of whole-cell catalysis of the *E. coli* system for hydrogen production. However, it can help to understand and characterize the H_2 producing machinery originated from different bacterial and cyanobacterial species *via* heterologous expression.

E. coli uses the FHL system for the production of H_2. This protein complex includes the enzyme products coded by genes formate dehydrogenase-H (*fdhF*) and hydrogenase 3 (*hycABDEFGHI*) (Bagramyan et al., 2003). A repressor protein HycA encoded by as gene *hycA* controls the expression of this complex in the cell. FhlA acts as an up-regulator of the FHL system a transcription factor. Apart from these, the H_2 production rate is controlled by the gene products which make availability of formate to FHL systems like formate dehydrogenase-N and formate dehydrogenase-O. In addition to these, the formate transporter FocA also plays a crucial role in the maintenance of sufficient formate to FHL and helps in the high production of H_2 by the cell (Rossmann et al., 1991, Andrews et al., 1997). The majority of the strategies in *E. coli* for strain engineering include the metabolic engineering to accumulate the formate in the cell, blocking the hydrogen uptake proteins, up-regulating the expression of FHL complex encoding genes. There are few efforts were also put forward to express non-native hydrogenases in *E. coli* to enhance the H_2 production.

The theoretical value of hydrogen productivity from one mol of glucose in *E. coli* cell is 2 mol of H_2 considering formate as the substrate to the FHL system. Experimentally reaching the theoretical value of H_2 production is not possible since the cell invests the energy for the biomass generation and other cellular processes. Hence, researchers look for the best possible outputs of H_2 production. In this context, metabolic engineering is the major aspect where the metabolic flux was diverted towards the accumulation of formate in the cell and feeding this formate to the FHL system. Since, FHL is the sole machinery in *E. coli*, any effort cutting down the metabolic pathways using formate other than feeding to FHL will enhance H_2 productivity. A significant contribution in this context is knockout studies make in *E. coli* by Maeda et al. (2012, 2018). The studies are resulted in reaching near to the theoretical values from glucose and formate as a carbon source (Maeda et al., 2007, 2008, 2012). To achieve the theoretical values selectively from the formate as the substrate in the feedstock; the authors used multiple strategies for engineering the strain to improved H_2 production. The strategies are as follows:

- Deleting the repressor of FHL expression (hycA) helped in the better expression of the FHL complex.
- Overexpression of gene *fhlA*, which encodes the transcription factor FhlA and assists in the up-regulation of the FHL complex.
- Inactivating the H_2 consuming hydrogenases 1 and 2 by knockout of *hycB* and *hycC*, which codes for the subunits responsible for the activity of hydrogenase 1 and 2, respectively.
- Deleting the gene *fdoG* helped in inactivating the FDH, which conducts the catalysis of converting formate to CO_2 without releasing H_2.

Glucose is considered the most versatile substrate for the metabolism of microorganisms. It is considered the starting moiety to generate formate and is an easily available carbon substrate many researchers considered it as a carbon substrate for their studies to produce H_2. The basic strategy using glucose as the carbon source is, majority of the studies consider diverting the metabolic flux towards formate accumulation, further using the best combination of the above-mentioned strategies to get the best yields from formate. Hence, *E. coli* with disabling hydrogenase 1 and 2 to block the uptake of H_2 by cell and knockout of *hycA* gene resulted in the enhancement of H_2 production. The basic strain with *hyaB*$^-$, *hyaC*$^-$ and *hyaA*$^-$ is considered for metabolic engineering for enhancing the H_2 productivity taking glucose as a carbon source. In addition, in a significant study where FhlA with N-terminal truncation protein upon overexpression, good improvement in H_2 production was observed in *E. coli* (Self et al., 2001).

The basic glucose metabolism through glycolysis generates the pyruvate from glucose. This will be converted to organic acids like formate, lactate or succinate in dark fermentation. Among these products, H_2 production will be benefitted if the flux is moved towards formate rather than other products. Hence, many studies considered this and upon diverting the metabolic flux towards formate resulted in a successful enhancement in H_2 productivity

(Yoshida et al., 2006, Maeda et al., 2007, Manish et al., 2007, Fan et al., 2009, Kim et al., 2009, Maeda et al., 2012). In these studies, selective genes like phosphoenolpyruvate carboxylase (*ppc*), formate reductase (*frd*ABCD), lactate dehydrogenase (*idh*A) were deleted for enhancing the productivity along with the deletions of gene coding for hydrogenase 1 and 2. In addition to studying the knockouts for diverting the metabolic flux, researchers also studied the overexpression of '*Fnr*', a global transcriptional regulator and found to enhance the H_2 production. Among the many strains developed, the best productivity was achieved with recombinant *E. coli* strain carrying the knockouts for the seven genes *hya*B, *hyb*C, *hyc*A, *fdo*G, *ldh*A, *frd*C, and *ace* (Maeda et al., 2012) or five genes knockout combination of *hya*AB, *hyb*ABC, *hyc*A, *ldh*A, *frd*BC/*hyc*A, *hya*, *hyb*, *ldhA*, and *frdAB* (Kim et al., 2009, Mathews et al., 2010) and three gene knockouts *hya*, *hyb*, and *ldhA* gave best H_2 productivity (Turcot et al., 2008).

Although glucose is a basic carbon source for the fermentation, using the low-cost carbon source indirectly can help in cutting down the final cost of the H_2 production. Hence, using the lignocellulosic sugars and crude glycerol is considered the best alternative carbon source for the fermentation. Though lignocellulosic biomass could not be used directly as a carbon source, at least in the case of *E. coli* fermentation, crude glycerol can be considered for the H_2 production. Crude glycerol is a waste biomass obtained as a by-product after the transesterification of vegetable oils during biodiesel production.

The benefit of using crude glycerol is, it could be used directly or after a minor process of purification, can be directly added to the medium as a carbon source for the *E. coli* growth, and successfully used whole-cell biocatalysis for chemical and fuel synthesis (Sudheer et al., 2018). The studies suggest that the physiological condition for the H_2 production is very diverse and not much correlated with the conditions of H_2 production when glucose is used as a carbon source (Dharmadi et al., 2006, Gonzalez et al., 2008). The studies demonstrated that the alkaline pH of the cell favored hydrogen production when glycerol is used as a carbon source and needs elevated concentrations of potassium and phosphate. Although multiple studies are available for the H_2 production using glycerol as a carbon source, still the mechanism and suitable conditions favorable for glycerol fermentation are far less than the fermentation in the presence of glucose. Moreover, the studies and results were contradictory and experimental yields were considerably different from each other.

Theoretically, the molar stoichiometry of the H_2 production from glycerol is 1 mol H_2 per 1 mol of glycerol. Moreover, using glycerol as a carbon source in anaerobic fermentation, the cell can have an extra $NADPH^+$, which can be used as an added cofactor, which could be used by the cell. Hence, the researchers view that, glycerol fermentation will be more beneficial and, if a good recombinant strain is developed, it could be a great success in the perspective of commercial production (Maeda et al., 2018). To date, the best study in developing recombinant *E. coli* for glycerol fermentation is by knockout of seven genes for the accumulation of formate. The selected gene knockout is fumarate reductase (*frdC*), formate dehydrogenase (*fdnG*), lactate dehydrogenase (*ldhA*), nitrate reductase (*narG*), phosphoenolpyruvate carboxylase (*ppc*), methylglyoxal synthase (*mgsA*) and the regulator of the transcriptional regulator FhlA (*hycA*) (Tran et al., 2015). Blocking of the methylglyoxal synthesis in the cell has a good effect of diverting the metabolic flux towards formate accumulation. The resulting strain was successfully able to reach near to the theoretical values (1 mole of H_2 form 1 mole of glycerol) (Maeda et al., 2018).

Although selective gene deletion of auxiliary pathways improved the formate accumulation and H_2 production in *E. coli*, understanding the pattern of genes/gene networks influencing in a genome level for the production of H_2 is still limited. Hence, Tran et al. applied the random mutagenesis for identifying the genes responsible for the H_2 productivity in *E. coli* (Tran et al., 2015, Tran et al., 2015). The individual mutation of four genes, namely *aroM, gatZ, ycgR, and yfgI* were identified as responsible for the enhancement of H_2 production, and these gene mutations resulted in enhancement of H_2 yields up to 1.6 folds. Moreover, the mutant strains also increased the growth rate under the fermentation in glycerol as the sole carbon source. Hu and Wood in their study, but efforts to isolate adoptive mutagenesis and successfully isolated a mutant strain HW2 which is holding the ability of H_2 production up to 20 times in glycerol medium and also have the better cell growth up to five-fold compared to the original parent strain BW25113 *frdC* (Hu et al., 2010). The transcriptome analysis of the mutant strain revealed that this strain is defective of fructose-1,6-bisphosphatase (encoded by *fbp*), formate transportation(*focA*), and tagatose-1,6-bisphosphate aldolase (*gatYZ*). These studies helped in gaining the knowledge of the gene networks, which influence the H_2 production in glycerol fermentation. However, there should be more studies need to be conducted at the genomic, transcriptomic, and proteomic level to understand the comprehensive factors which influence the H_2 production in the cell.

Apart from the studies on a knockout, random mutagenesis, and overexpression of the genes, few studies were also conducted in *E. coli* for heterologous expression of genes for the enhancement and/or for the characterization of the genes related to hydrogen production from the other organism. In this context, various hydrogenases from diverse organisms were tried to express in *E. coli.* Significant among these are the hydrogenases from *Enterobacter cloacae* (Mishra et al., 2004, Chittibabu et al., 2006), *Ethanoligenens harbinense* (Zhao et al., 2010), *Rhodobacter sphaeroides* (Lee et al., 2010), *Clostridium acetobutylicum* and *Clostridium pasteurianum* (Akhtar et al., 2008, 2009) were heterologously expressed in *E. coli* BL21 which is known to have lack of ability to produce hydrogen. Apart from expressing the hydrogenases, few more researchers also expressed the genes related to substrate transportation, substrate utilization, key metabolic genes that could enhance the accumulation of H_2.

The significant among those studies which resulted in enhancement of H_2 productivity in *E. coli* were the expression of *scrB* (encode β-D-fructofuranoside fructohydrolase catalyzes the hydrolysis of sucrose 6-phosphate to β-D-fructose and α-D-glucose 6-phosphate) and *scrR* (encodes the negative repressor of the scr regulon). *E. coli* being the model organism, with many advantages of its molecular understanding, it has all potentials to become an efficient microbial cell factory for future biohydrogen production.

6.4 CONCLUSIONS

Hydrogen fuel is the only zero-emission fuel and also a next-generation green fuel, which can address the environmental issue of global carbon emissions. To make this fuel to replace the fossil fuel for transportation and industry utilities, more innovative methods are required to be designed and established for establishing industrial-scale production. The important points to be considered for making hydrogen as the best alternative fuel, the points to be considered are the following:

- Technologies to be developed to use the industrial and domestic waste and/or effluents for the production of H_2.
- Efficient strains suitable for each context of bio-waste utilization need to be isolated and characterized.
- Metabolism and molecular process of H_2 production should be studied and understand well.
- More dynamic recombinant strains should be made by using or inventing novel bio-engineering/molecular engineering/cell system engineering technologies.

Moreover, a feasible reactor system for the production of H_2 also needs to be designed for efficient production at the commercial level. Providing the platform for joining the hands of government-private organizations for the research and enhancing the funding by government and industries will allow more studies in all aspects above discussed will certainly lead H_2 fuel originating as next-generation clean fuel, which is free of carbon emissions, and keep the environment safe.

ACKNOWLEDGMENTS

The authors would like to thank the Department of Biotechnology (DBT), Government of India for the funding under Ramalingaswami re-entry fellowship (Project # AUR002). The authors also thank Prof. Rajendra Kumar Pandey (Vice-Chancellor, Amity University Chhattisgarh, India), Dr. Ravi Kanth Singh (Director, Amity Institute of Biotechnology) for their kind support and encouragement.

KEYWORDS

- **biohydrogen**
- **cyanobacteria**
- **electron transport chain**
- ***Escherichia coli***
- **ferredoxin**
- **genetic engineering**
- **hydrogenases**
- **O_2 evolving complex**
- **photosystem-II**
- **phycocyanin**
- **phycoerythrin**
- **plastoquinone**
- **reaction center-1**

REFERENCES

Akhtar, M. K., & Jones, P. R., (2008). Deletion of iscR stimulates recombinant clostridial Fe-Fe hydrogenase activity and H_2-accumulation in *Escherichia coli* BL21(DE3). *Appl. Microbiol. Biotechnol., 78*, 853–862.

Akhtar, M. K., & Jones, P. R., (2009). Construction of a synthetic YdbK-dependent pyruvate:H_2 pathway in *Escherichia coli* BL21(DE3). *Metab. Eng., 11*, 139–147.

Andrews, S. C., Berks, B. C., McClay, J., Ambler, A., Quail, M. A., Golby, P., & Guest, J. R., (1997). A 12-cistron *Escherichia coli* operon (hyf) encoding a putative proton-translocating formate hydrogenlyase system. *Microbiology, 143*, 3633–3647.

Asada, Y., Koike, Y., Schnackenberg, J., Miyake, M., Uemura, I., & Miyake, J., (2000). Heterologous expression of clostridial hydrogenase in the *Cyanobacterium Synechococcus* PCC7942 *Biochim. Biophys. Acta, 1490*, 269–278.

Baebprasert, W., Jantaro, S., Khetkorn, W., Lindblad, P., & Incharoensakdi, A., (2011). Increased H_2 production in the cyanobacterium *Synechocystis* sp. strain PCC 6803 by redirecting the electron supply via genetic engineering of the nitrate assimilation pathway. *Metab. Eng., 13*, 610–616.

Bagramyan, K., & Trchounian, A., (2003). Structural and functional features of formate hydrogen lyase, an enzyme of mixed-acid fermentation from *Escherichia coli. Biochemistry, 68*, 1159–1170.

Bandyopadhyay, A., Stöckel, J., Min, H., Sherman, L. A., & Pakrasi H. B., (2010). High rates of photobiological H_2 production by a cyanobacterium under aerobic conditions. *Nature Comm., 1*, 139.

Barber, J., (2008). Crystal structure of the oxygen-evolving complex of photosystem II. *Inorganic Chem., 47*, 1700–1710.

Bonner, M., Botts, T., McBreen, J., Mezzina, A., Salzano, F., & Yang, C., (1984). Status of advanced electrolytic hydrogen production in the United States and abroad. *Int. J. Hydrogen Energy, 9*, 269–275.

Brosseau, J. D., & Zajic, J. E., (1982). Hydrogen-gas production with *Citrobacter intermedim* and *Clostridium pasteurianum. J. Chem. Technol. Biotechnol., 32*, 496–502.

Bryant, D. A., & Frigaard, N. U., (2006). Prokaryotic photosynthesis and phototrophy illuminated. *Trends in Microbiol., 14*, 488–496.

Burrows, E. H., Chaplen, F. W., & Ely, R. L., (2011). Effects of selected electron transport chain inhibitors on 24-h hydrogen production by *Synechocystis* sp. PCC 6803. *Bioresour. Technol., 102*, 3062–3070.

Chattanathan, S. A., Adhikari, S., & Abdoulmoumine, N., (2012). A review on the current status of hydrogen production from bio-oil. *Renew. Sust. Energy Rev., 16*, 2366–2372.

Chittibabu, G., Nath, K., & Das, D., (2006). Feasibility studies on the fermentative hydrogen production by recombinant *Escherichia coli* BL-21. *Process Biochem., 41*, 682–688.

Cooley, J. W., & Vermaas, W. F. J., (2001). Succinate dehydrogenase and other respiratory pathways in thylakoid membranes of *Synechocystis* sp. Strain PCC 6803: Capacity comparisons and physiological function. *J. Bacteriol., 183*, 4251–4258.

Cournac, L., Guedeney, G., Peltier, G., & Vignais, P. M., (2004). Sustained photo evolution of molecular hydrogen in a mutant of *Synechocystis* sp. strain PCC 6803 deficient in the type I NADPH-dehydrogenase complex. *J. Bacteriol., 186*, 1737–1746.

Dharmadi, Y., Murarka, A., & Gonzalez, R., (2006). Anaerobic fermentation of glycerol by *Escherichia coli*: A new platform for metabolic engineering. *Biotechnol. Bioeng., 94*, 821–829.

Ducat, D. C., Sachdeva, G., & Silver, P. A., (2011). Rewiring hydrogenase-dependent redox circuits in cyanobacteria. *PNAS, 108*, 3941–3946.

Fan, Z., Yuan, L., & Chatterjee, R., (2009). Increased hydrogen production by genetic engineering of *Escherichia coli. PloS One, 4*, e4432.

Genç, M. S., Çelik, M., & Karasu, İ., (2012). A review on wind energy and wind-hydrogen production in Turkey: A case study of hydrogen production via electrolysis system supplied by wind energy conversion system in Central Anatolian Turkey. *Renew. Sust. Energy Rev., 16*, 6631–6646.

Gonzalez, R., Murarka, A., Dharmadi, Y., & Yazdani, S. S., (2008). A new model for the anaerobic fermentation of glycerol in enteric bacteria: Trunk and auxiliary pathways in *Escherichia coli. Metab. Eng., 10*, 234–245.

Grossman, A. R., Schaefer, M. R., Chiang, G. G., & Collier, J. L., (1993). The phycobilisome, a light-harvesting complex responsive to environmental conditions. *Microbiol. Rev., 57*, 725–749.

Gutekunst, K., Hoffmann, D., Westernströer, U., Schulz, R., Garbe-Schönberg, D., & Appel, J., (2018). In-vivo turnover frequency of the cyanobacterial NiFe-hydrogenase during photohydrogen production outperforms *in-vitro* systems. *Sci. Rep., 8*, 6083.

Hallenbeck, P. C., & Ghosh, D., (2009). Advances in fermentative biohydrogen production: The way forward? *Trends Biotechnol., 27*, 287–297.

Happe, T., Schütz, K., & Böhme, H., (2000). Transcriptional and mutational analysis of the uptake hydrogenase of the filamentous cyanobacterium *Anabaena variabilis* ATCC 29413. *J. Bacteriol., 182*, 1624–1631.

Heap, J. T., Kuehne, S. A., Ehsaan, M., Cartman, S. T., Cooksley, C. M., Scott, J. C., & Minton, N. P., (2010). The ClosTron: Mutagenesis in clostridium refined and streamlined. *J. Microbiol. Methods, 80*, 49–55.

Heap, J. T., Pennington, O. J., Cartman, S. T., Carter, G. P., & Minton, N. P., (2007). The ClosTron: A universal gene knock-out system for the genus *Clostridium. J. Microbiol. Methods, 70*, 452–464.

Heidorn, T., Camsund, D., Huang, H. H., Lindberg, P., Oliveira, P., Stensjö, K., & Lindblad, P., (2011). Synthetic biology in cyanobacteria engineering and analyzing novel functions. *Methods Enzymol., 497*, 539–579.

Hu, H., & Wood, T. K., (2010). An evolved *Escherichia coli* strain for producing hydrogen and ethanol from glycerol. *Biochem. Biophys. Res. Comm., 391*, 1033–1038.

Huang, H. H., & Lindblad, P., (2013). Wide-dynamic-range promoters engineered for *Cyanobacteria. J. Biol. Eng., 7*, 10.

Igarashi, R. Y., & Seefeldt, L. C., (2003). Nitrogen fixation: The mechanism of the Mo-dependent nitrogenase. *Crit. Rev. Biochem. Mol. Biol., 38*, 351–384.

Kern, J., & Renger, G., (2007). Photosystem II: Structure and mechanism of the water: Plastoquinone oxidoreductase. *Photosynthesis Res., 94*, 183–202.

Khan, A. Z., & Bilal, M., (2019). State-of-the-art genetic modalities to engineer *Cyanobacteria* for sustainable biosynthesis of biofuel and fine-chemicals to meet bio-economy challenges. *Life, 9*, 54.

Khetkorn, W., Baebprasert, W., Lindblad, P., & Incharoensakdi, A., (2012). Redirecting the electron flow towards the nitrogenase and bidirectional Hox-hydrogenase by using specific inhibitors results in enhanced H_2 production in the cyanobacterium *Anabaena siamensis* TISTR 8012. *Bioresour. Technol., 118*, 265–271.

Khetkorn, W., Khanna, N., Incharoensakdi, A., & Lindblad, P., (2013). Metabolic and genetic engineering of cyanobacteria for enhanced hydrogen production. *Biofuels, 4*, 535–561.

Khetkorn, W., Lindblad, P., & Incharoensakdi, A., (2012). Inactivation of uptake hydrogenase leads to enhanced and sustained hydrogen production with high nitrogenase activity under high light exposure in the cyanobacterium *Anabaena siamensis* TISTR 8012. *J. Biol. Eng., 6*, 19.

Kim, S., Seol, E., Oh, Y. K., Wang, G. Y., & Park, S., (2009). Hydrogen production and metabolic flux analysis of metabolically engineered *Escherichia coli* strains. *Int. J. Hydrogen Energy, 34*, 7417–7427.

Knoot, C. J., Ungerer, J., Wangikar, P. P., & Pakrasi, H. B., (2018). Cyanobacteria: Promising biocatalysts for sustainable chemical production. *J. Biol. Chem., 293*, 5044–5052.

Kovács, K. L., Maróti, G., & Rákhely, G., (2006). A novel approach for biohydrogen production. *Int. J. Hydrogen Energy, 31*, 1460–1468.

Lee, S. Y., Lee, H. J., Park, J. M., Lee, J. H., Park, J. S., Shin, H. S., Kim, Y. H., & Min, J., (2010). Bacterial hydrogen production in recombinant *Escherichia coli* harboring a HupSL hydrogenase isolated from *Rhodobacter sphaeroides* under anaerobic dark culture. *Int. J. Hydrogen Energy, 35*, 1112–1116.

Levin, D. B., & Chahine, R., (2010). Challenges for renewable hydrogen production from biomass. *Int. J. Hydrogen Energy, 35*, 4962–4969.

Lindberg, P., Devine, E., Stensjö, K., & Lindblad, P., (2012). HupW protease specifically required for processing of the catalytic subunit of the uptake hydrogenase in the cyanobacterium *Nostoc* sp. strain PCC 7120. *Appl. Environ. Microbiol., 78*, 273–276.

Lindberg, P., Schütz, K., Happe, T., & Lindblad, P., (2002). A hydrogen-producing, hydrogenase-free mutant strain of *Nostoc punctiforme* ATCC 29133. *Int. J. Hydrogen Energy, 27*, 1291–1296.

Lu, Y., Zhao, H., Zhang, C., Lai, Q., Wu, X., & Xing, X. H., (2010). Alteration of hydrogen metabolism of ldh-deleted *Enterobacter aerogenes* by overexpression of NAD(+)-dependent formate dehydrogenase. *Appl. Microbiol. Biotechnol., 86*, 255–262.

Madigan, M., & Jung, D., (2008). An overview of purple bacteria: Systematics, physiology, and habitats. In: Hunter, C. N., Daldal, F., Thurnauer, M. C., & Beatty, J. T., (eds.), *The Purple Phototrophic Bacteria* (pp. 1–15). Springer Nature.

Maeda, T., Sanchez-Torres, V., & Wood, T. K., (2007). Enhanced hydrogen production from glucose by metabolically engineered *Escherichia coli*. *Appl. Microbiol. Biotechnol., 77*, 879–890.

Maeda, T., Sanchez-Torres, V., & Wood, T. K., (2007). *Escherichia coli* hydrogenase 3 is a reversible enzyme possessing hydrogen uptake and synthesis activities. *Appl. Microbiol. Biotechnol., 76*, 1035–1042.

Maeda, T., Sanchez-Torres, V., & Wood, T. K., (2008). Metabolic engineering to enhance bacterial hydrogen production. *Microb. Biotechnol., 1*, 30–39.

Maeda, T., Sanchez-Torres, V., & Wood, T. K., (2012). Hydrogen production by recombinant *Escherichia coli* strains. *Microb. Biotechnol., 5*, 214–225.

Maeda, T., Tran, K. T., Yamasaki, R., & Wood, T. K., (2018). Current state and perspectives in hydrogen production by *Escherichia coli*: Roles of hydrogenases in glucose or glycerol metabolism. *Appl. Microbiol. Biotechnol., 102*, 2041–2050.

Manish, S., Venkatesh, K. V., & Banerjee, R., (2007). Metabolic flux analysis of biological hydrogen production by *Escherichia coli*. *Int. J. Hydrogen Energy, 32*, 3820–3830.

Marcelo, D., & Dell'Era, A., (2008). Economical electrolyser solution. *Int. J. Hydrogen Energy, 33*, 3041–3044.

Masukawa, H., Inoue, K., Sakurai, H., Wolk, C. P., & Hausinger, R. P., (2010). Site-directed mutagenesis of the *Anabaena* sp. strain PCC 7120 nitrogenase active site to increase photobiological hydrogen production. *Appl. Environ. Microbiol., 76*, 6741–6750.

Mathews, J., Li, Q., & Wang, G., (2010). Characterization of hydrogen production by engineered Escherichia coli strains using rich defined media. *Biotechnol. Bioproc. Eng., 15*, 686–695.

Mishra, J., Khurana, S., Kumar, N., Ghosh, A. K., & Das, D., (2004). Molecular cloning, characterization, and overexpression of a novel [Fe]-hydrogenase isolated from a high rate of hydrogen-producing *Enterobacter cloacae* IIT-BT 08. *Biochem. Biophys. Res. Comm., 324*, 679–685.

Mujeebu, M. A., (2016). Hydrogen and syngas production by super adiabatic combustion: A review *Appl. Energy, 173*, 210–224.

Nakashimada, Y., Rachman, M. A., Kakizono, T., & Nishio, N., (2002). Hydrogen production of *Enterobacter aerogenes* altered by extracellular and intracellular redox states. *Int. J. Hydrogen Energy, 27*, 1399–1405.

Nanda, S., Rana, R., Zheng, Y., Kozinski, J. A., & Dalai, A. K., (2017). Insights on pathways for hydrogen generation from ethanol. *Sustain. Energy Fuels, 1*, 1232–1245.

Oh, Y. K., Raj, S. M., Jung, G. Y., & Park, S., (2013). Chapter 3 - Metabolic engineering of microorganisms for biohydrogen production. In: Pandey, A., Chang, J. S., Hallenbecka, P. C., & Larroche, C., (eds.), *Biohydrogen* (pp. 45–65). Amsterdam, Elsevier.

Okolie, J. A., Nanda, S., Dalai, A. K., Berruti, F., & Kozinski, J. A., (2020). A review on subcritical and supercritical water gasification of biogenic, polymeric and petroleum wastes to hydrogen-rich synthesis gas. *Renew. Sust. Energy Rev., 119*, 109546.

Okolie, J. A., Rana, R., Nanda, S., Dalai, A. K., & Kozinski, J. A., (2019). Supercritical water gasification of biomass: A state-of-the-art review of process parameters, reaction mechanisms and catalysis. *Sustain. Energy Fuels, 3*, 578–598.

Rossmann, R., Sawers, G., & Böck, A., (1991). Mechanism of regulation of the formate-hydrogenlyase pathway by oxygen, nitrate, and pH: Definition of the formate regulon. *Mol. Microbiol., 5*, 2807–2814.

Saga, Y., Shibata, Y., & Tamiaki, H., (2010). Spectral properties of single light-harvesting complexes in bacterial photosynthesis. *J. Photochem. Photobiol. C: Photochem. Rev., 11*, 15–24.

Sarangi, P. K., & Nanda, S., (2020). Biohydrogen production through dark fermentation. *Chem. Eng. Technol., 43*, 601–612.

Schütz, K., Happe, T., Troshina, O., Lindblad, P., Leitão, E., Oliveira, P., & Tamagnini, P., (2004). Cyanobacterial H_2 production — a comparative analysis. *Planta, 218*, 350–359.

Self, W. T., Hasona, A., & Shanmugam, K. T., (2001). N-terminal truncations in the FhlA protein result in formate- and MoeA-independent expression of the hyc (formate hydrogenlyase) operon of *Escherichia coli*. *Microbiology, 147*, 3093–3104.

Singh, S., Kumar, R., Setiabudi, H. D., Nanda, S., & Vo, D. V. N., (2018). Advanced synthesis strategies of mesoporous SBA-15 supported catalysts for catalytic reforming applications: A state-of-the-art review. *Appl. Catal. A: Gen., 559*, 57–74.

Sinha, P., & Pandey, A., (2011). An evaluative report and challenges for fermentative biohydrogen production. *Int. J. Hydrogen Energy, 36*, 7460–7478.

Skizim, N. J., Ananyev, G. M., Krishnan, A., & Dismukes, G. C., (2012). Metabolic pathways for photobiological hydrogen production by nitrogenase- and hydrogenase-containing unicellular cyanobacteria *Cyanothece*. *J. Biol. Chem., 287*, 2777–2786.

Stephanopoulos, G., (2007). Challenges in engineering microbes for biofuels production. *Science, 315*, 801–804.

Stojić, D. L., Marčeta, M. P., Sovilj, S. P., & Miljanić, Š. S., (2003). Hydrogen generation from water electrolysis—possibilities of energy saving. *J. Power Sources, 118*, 315–319.

Sudheer, P., Seo, D., Kim, E. J., Chauhan, S., Chunawala, J. R., & Choi, K. Y., (2018). Production of (Z)-11-(heptanoyloxy)undec-9-enoic acid from ricinoleic acid by using crude glycerol as sole carbon source in engineered *Escherichia coli* expressing BVMO-ADH-FadL. *Enzyme Microb. Technol., 119*, 45–51.

Tapia-Venegas, E., Ramirez-Morales, J. E., Silva-Illanes, F., Toledo-Alarcón, J., Paillet, F., Escudie, R., Lay, C. H., et al., (2015). Biohydrogen production by dark fermentation: Scaling-up and technologies integration for a sustainable system. *Rev. Environ. Sci. Bio/Technol., 14*, 761–785.

Tran, K. T., Maeda, T., Sanchez-Torres, V., & Wood, T. K., (2015). Beneficial knockouts in *Escherichia coli* for producing hydrogen from glycerol. *Appl. Microbiol. Biotechnol., 99*, 2573–2581.

Turcot, J., Bisaillon, A., & Hallenbeck, P. C., (2008). Hydrogen production by continuous cultures of *Escherichia coli* under different nutrient regimes. *Int. J. Hydrogen Energy, 33*, 1465–1470.

Turner, J. A., (2004). Sustainable hydrogen production. *Science, 305*, 972.

Utgikar, V., & Thiesen, T., (2006). Life cycle assessment of high temperature electrolysis for hydrogen production via nuclear energy. *Int. J. Hydrogen Energy, 31*, 939–944.

Weissman, J. C., & Benemann, J., (1977). Hydrogen production by nitrogen starved cultures of *Anabaena cylindrical. Appl. Environ. Microbiol., 33*, 123–131.

Weyman, P. D., Vargas, W. A., Tong, Y., Yu, J., Maness, P. C., Smith, H. O., & Xu, Q., (2011). Heterologous expression of *Alteromonas macleodii* and *Thiocapsa roseopersicina* [NiFe] hydrogenases in *Synechococcus elongates*. *PloS One, 6*, e20126.

Yoshida, A., Nishimura, T., Kawaguchi, H., Inui, M., & Yukawa, H., (2006). Enhanced hydrogen production from glucose using ldh- and frd-inactivated *Escherichia coli* strains. *Appl. Microbiol. Biotechnol., 73*, 67–72.

Yoshino, F., Ikeda, H., Masukawa, H., & Sakurai, H., (2007). High photobiological hydrogen production activity of a *Nostoc* sp. PCC 7422 uptake hydrogenase-deficient mutant with high nitrogenase activity. *Marine Biotechnol., 9*, 101–112.

Zhang, C., Ma, K., & Xing, X. H., (2009). Regulation of hydrogen production by *Enterobacter aerogenes* by external NADH and NAD+. *Int. J. Hydrogen Energy, 34*, 1226–1232.

Zhang, X., Sherman, D. M., & Sherman, L. A., (2014). The uptake hydrogenase in the unicellular diazotrophic cyanobacterium *Cyanothece* sp. Strain PCC 7822 protects nitrogenase from oxygen toxicity. *J. Bacteriol., 196*, 840.

Zhao, H., Ma, K., Lu, Y., Zhang, C., Wang, L., & Xing, X. H., (2009). Cloning and knockout of formate hydrogen lyase and H_2-uptake hydrogenase genes in *Enterobacter aerogenes* for enhanced hydrogen production. *Int. J. Hydrogen Energy, 34*, 186–194.

Zhao, X., Xing, D., Zhang, L., & Ren, N., (2010). Characterization and overexpression of a [FeFe]-hydrogenase gene of a novel hydrogen-producing bacterium *Ethanoligenens harbinense*. *Int. J. Hydrogen Energy, 35*, 9598–9602.

CHAPTER 7

Hydrogen Production through Microbial Electrolysis

LATIKA BHATIA,[1] PRAKASH K. SARANGI,[2] and SONIL NANDA[3]

[1]*Department of Microbiology and Bioinformatics, Atal Bihari Vajpayee University, Bilaspur, Chhattisgarh, India*
E-mail: latikabhatia1@yahoo.co.uk (Latika Bhatia)

[2]*Directorate of Research, Central Agricultural University, Imphal, Manipur, India*

[3]*Department of Chemical and Biological Engineering, University of Saskatchewan, Saskatoon, Saskatchewan, Canada*

ABSTRACT

The production of biological hydrogen is one of the cleanest energy processing routes. Substantial research is underway to enhance the production rate of hydrogen, which includes improving reactor configurations, unit operations, genetic modification of microorganisms, metabolic engineering, developing membrane-less bioreactors, and using novel and low-cost organic feedstocks. These constrain needs to be addressed for the effective commercialization of hydrogen production. Appropriate microbial cultures are required to manage waste matter proficiently, which are frequently complex. Electrohydrogenesis is a process for producing hydrogen in microbial electrolysis cells at better yields with greater competence. This is an efficient solution to resolve two issues viz. the generation of clean energy and bioremediation of waste material.

7.1 INTRODUCTION

Rapid industrial development and urban expansion have posed a thoughtful issue of energy crisis both for developing and developed nations (Nanda et al., 2014). Non-renewable sources of energy, whether it is fossil energy or nuclear energy, has contended the global requirement of energy need from the past many decades and are still sufficing this need. Serious issues are to be addressed by many nations in the term of sustainability and other environmental issues like global warming and pollution generated from the usage of fossil fuels (Nanda et al., 2015, 2016; Sarangi and Nanda, 2018, 2019). Pollution is also contributed by the accumulation of anthropogenic wastes including non-biodegradable petrochemical derivatives, microplastics, and non-biodegradable fractions of municipal solid wastes that needs attention to be recycled sustainably (Nanda and Berruti, 2021a,b,c). The waste-to-energy (thermochemical or biological transformations) approach is a feasible strategy to manage such wastes and generate clean energy (Winfield et al., 2016). The need for an hour demands the further exploration of novel renewable energy technologies to suffice global energy demands while addressing the environmental issues (Bhatia et al., 2020; Sarangi et al., 2017, 2019).

The world has started exploring many forms of renewable energy, whether it is solar energy, wind energy, tidal energy, geothermal energy, and discarded biomass (e.g., lignocellulosic biomass, cattle dung, industrial waste matters, food waste, waste cookery oil, etc.), (Nanda et al., 2019, 2020; Sarangi and Nanda, 2020; Sarangi et al., 2020; Okolie et al., 2020). Microbial electrolysis cells are recent options being explored as a renewable source of energy because of the zero emissions of greenhouse gases (Rahimnejad et al., 2015).

7.2 HYDROGEN AS A FUEL

Hydrogen is a promising future fuel as its energy content is high (121 MJ/kg) and its carbon emissions are zero (Nanda et al., 2017). The energy density is high in the case of hydrogen when compared with other fuels. Space shuttles employ liquid hydrogen as a fuel. As the flammability of hydrogen is high, this property of hydrogen makes its storage a difficult task. The storage of hydrogen demands a cryogenic tank as its boiling point is around –252.9°C. These storage tanks are properly insulated to prevent the vulnerability of boil-off (Holladay et al., 2009). Fuel cells use hydrogen as a fuel to generate power. Hydrogen finds its utility as a fuel for a fuel cell or direct combustion in an internal combustion engine (Bielen et al., 2013). Hydrogen supports the synthesis of chemicals and electrical generation and storage with the fuel cells, thereby proving itself as an energy carrier and vector. Hydrogen can generate heat and water upon combustion, which is of high relevance to industries and sustainability (Hallenbeck and Ghosh, 2009). Many initiatives are underway to develop hydrogen-based vehicles and catalytic incineration schemes for its utilization in industries, automobiles, households as well as hydrogen-supported small power-producing plants.

Hydrogen serves as a crucial feedstock for the chemical industries (Okolie et al., 2019). Presently, the processing of fossil fuels suffices the prerequisite of hydrogen for many processes such as hydrotreating, hydroprocessing, reforming, Fischer-Tropsch synthesis and fuel upgrading (Singh et al., 2018; Reddy et al., 2020). Hydrogen production through these routes has huge greenhouse gas footprints (Lee et al., 2010). On the other hand, hydrogen production techniques through biological routes are not only sustainable but carbon-neutral (Redwood et al., 2009). Agro-residues, lignocellulosic material, food processing waste, water plants, algae, and municipal solid waste are examples of some potential biomasses, which can be employed for hydrogen production through biological pathways (Rupprecht et al., 2006; Beer et al., 2009). Many environmental, technical, and socio-economic benefits are associated with hydrogen that makes it potential in the sequence of fuel evolution (Das and Veziroglu, 2008).

7.3 MICROBIAL ROUTES FOR HYDROGEN PRODUCTION

There are many routes for the formation of hydrogen. The production of hydrogen involving a biological system is far a better approach as compared to that of the other thermochemical methods (e.g., reforming, gasification, pyrolysis, photo-electrochemical, photocatalytic, etc.). In the biological methods for hydrogen production, microbial metabolism generates hydrogen at ambient temperatures and pressures with no toxic byproduct formation, which can pose a major risk to the environment. A wide diversity of microorganisms have the potential to generate hydrogen. These microorganisms use a broad spectrum of substrates for fermentative biohydrogen production. The availability of these renewable substrates makes this an efficient, sustainable, economic, and eco-friendly approach of energy generation in the form of hydrogen.

Small-scale installations can employ these decentralized energy production techniques in places where the availability of biomass or wastes is substantial, and that could save costs relating to biomass transport and drying. Several microorganisms have the potential to generate hydrogen that includes algae and bacteria (Nandi and Sengupta, 1998). These microorganisms are metabolically different but require unique substrates to function under particular process conditions (Das and Veziroglu, 2001).

The photofermentation requires light energy to produce hydrogen. The dark fermentation employs protons to accept electrons to produce hydrogen (Lee et al., 2010). This process supports the effective management of municipal solid wastes or other organic wastes, which can act as substrates for hydrogen generation through microbial decomposition. This approach has a better hydrogen yield when compared to the other biological methods of biohydrogen production. The fermentation process is often coupled with other pathways such as microbial electrolysis cells to enhance the proportion of substrate converted to biohydrogen. Some recent research indicates that the amalgamation of dark fermentation with microbial electrolysis cell can advance biohydrogen yields (Lee et al., 2010).

Microbial electrochemical systems (MES) are the novel technologies applied in various sectors such as:

- Microbial fuel cells (MFC) for combined applications for wastewater treatment and electricity production
- Microbial electrolysis cells (MEC) for biohydrogen and other platform chemical production
- Microbial dialysis cells (MDC) for water desalination.

Microbial electrolysis is a newly developed technology where the organic wastes are converted to hydrogen. It depends on two energy sources, i.e., bacteria mediated oxidization of organic matter and electricity. The reactors employed for this procedure are known as microbial electrolysis cells (MEC). Microbial electrolysis cells are the reactors that generate hydrogen by the incorporation of electrical energy. This process is known as electrohydrogenesis. Working of MEC can be partially related to the working of microbial fuel cells where the potential of the anode (EAn) is usually around −0.3 V (Liu and Logan, 2004 Logan et al., 2006). Many biodegradable organic substrates can be employed in the operation of MECs that when oxidized by bacteria, they could generate electron and protons (Zuo et al., 2007).

Protons reach the cathode through the proton exchange membrane (separating anode and cathode). At the cathode, electrons cannot spontaneously combine with a proton to produce hydrogen. This becomes possible only when the potential of the cathode (ECat) is at least −0.414 V under normal biological conditions (i.e., temperature: 25°C, pH: 7 and hydrogen partial pressure: 1 bar). The reaction is non-spontaneous due to the negative cell voltage. When the voltage is incorporated more than 0.114 V, it favors the combination of electron and proton at the cathode to form hydrogen gas. To accomplish measurable production rates, overpotentials (the difference between the theoretical and actual potential) of 0.2 V or more must be applied at the anode and cathode (Cheng and Logan, 2007). A catalyst is needed at the cathode that supports the reduction of protons to electrons. This improves the kinetics of the reaction along with the addition of voltage to create a complementary reaction. This procedure of supplementing bacterial oxidation of organic material with a contribution of electrical energy is termed as electrohydrogenesis or biocatalyzed electrolysis (Figure 7.1) (Rozendal et al., 2006).

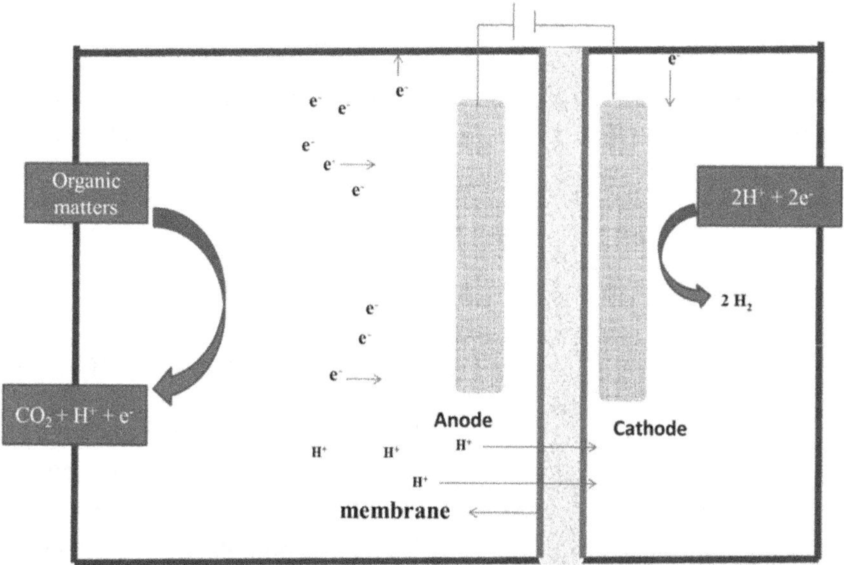

FIGURE 7.1 Schematic of a typical microbial electrolysis cell assembly.

Spontaneous production of hydrogen without the need of applying electrical energy is feasible if a suitable substrate (i.e., simple sugar, for example, glucose) is used. When this potential is combined with cathode potential, a positive value is obtained that indicates the spontaneous reaction for hydrogen formation. The cathode is an anaerobic chamber in MECs, unlike that of MFCs. Glucose can be fermented under anaerobic conditions by bacteria in the anode chamber. The products of fermentation such as acetate and butyrate usually do not support the spontaneous conversion to hydrogen.

It is important to understand the basic differences between MFCs and MECs. Like MFCs, MECs can also use a wide spectrum of organic material that ranges from agro-residues to wastewater. This organic matter can be biodegraded with the aid of microorganisms (Min et al., 2005; Zuo et al., 2006; Schubert, 2006). MECs can support the treatment of domestic wastewater (Escapa et al., 2014). Microorganisms and microbial consortia are employed in the anode chamber of the MFC where they oxide the wastewater or the pollutant, thereby generating the electrons and protons. These microorganisms for their metabolism of energy generation use electrons by electron transfer chain after which they travel to the cathode chamber completing the circuit. The protons generated in the anode chamber during the oxidation process travel via the proton exchange membrane (PEM) and reache the cathode. The cathode chamber is oxygenated where electrons and protons reduce the oxygen to produce water molecules, thereby completing the charge balance. In this process, chemical energy is converted to electrical energy (Osman et al., 2010). Extracellular electron transport mechanism of electrochemically active bacteria (EAB) support the transport of electrons, thereby generating voltage (Wang et al., 2014). MECs are similar to MFCs in many aspects but differ in having a cathode that is sealed to eliminate oxygen and supplementary participation of electricity to generate hydrogen gas at the cathode.

7.4 NOTABLE RESEARCH WORKS ON MICROBIAL ELECTROLYSIS CELLS

Using acetate as a substrate in a two-bottle system, Liu et al. (2005) reported hydrogen recoveries of more than 90% with low electrical energy requirement (0.5 kWh/m^3-H_2 at Eap of 0.25 V), which was striking in comparison to traditional water electrolysis (4.5–5 kWh/m^3-H_2). Many fermentation products have been tested for their efficiency as a substrate in MECs, chief among them include butyric, lactic, propionic, and valeric acid. Hydrogen recovered from these substrates ranges from 67–89% (Cheng and Logan, 2007). The biological cathode in the form of bacterial hydrogenase enzymes was also tested on the cathode side to see its impact on hydrogen recovery. The hydrogenase enzyme is sensitive towards oxygen and depicts low oxygen tolerance. It was found that with such an arrangement, hydrogen recovered was roughly about 49% (Rozendal et al., 2008).

Most of the studies on biological hydrogen production incorporated a membrane that separated anode from the cathode. The prominent reason behind the use of a membrane is to obtain the hydrogen of the highest purity and to restrain bacteria on the cathode to use hydrogen (Ditzig et al., 2007). This membrane is known as a proton exchange membrane, e.g., Nafion. It has functional groups such as $–SO^3$ that only allow liberated protons (H^+) to move through them (Kim et al., 2007). This membrane can potentially transfer other cationic species (Na^+, K^+, NH_4^+, Ca^{2+} and Mg^{2+}). Wastewater has these ions, 4–5 times more in concentration than H^+ ions (Zhao et al., 2006). There should be a satisfactory balance of protons in anode and cathode of MECs as the free protons are produced in the anode as a result of oxidation and are removed in the cathode due to reduction. If the cationic species move through the membrane to maintain electroneutrality between anode and cathode, other than the proton, it results in subsequent anode acidification and cathode alkalization (Harnisch et al., 2008). This pH gradient developed is responsible for performance losses in MECs. According to the Nernst equation, a unit alteration in pH adds to a potential loss of 0.06 V (Rozendal et al., 2007). Using a MEC with the Nafion membrane, Rozendal et al. (2007) illustrated a pH hike of 6.4, which matched to a 0.38 V loss of the applied 1.0 V.

Anion exchange membranes (AEMs) are also tested in MECs. The performance of AEM was found better in terms of power production when compared to Nafion, a cationic exchange membrane (CEM), and an ultrafiltration membrane (Kim et al., 2007). Ammonium ($–NH^{3+}$) functional groups are found on AEM that permits negatively charged phosphate species (HPO_4^{2-} and $H_2PO_4^-$) to diffuse through them. The movement of these ions across the membrane not only helps to buffer the pH drop in the anodic chamber but also to maintain the electroneutrality in anode and cathode chambers. The AEM in this way becomes responsible for the largest hike in pH in the cathode chamber that attributes to a loss of 0.26 V of the 1.0 V applied (Rozendal et al., 2007). MECs design should be such a way that pH gradient-associated potential losses is eliminated. Substantial loss occurs due to the diffusion of gases like H_2, CO_2, and CH_4 from the cathode to the anode chamber. This was predicted by Rozendal et al. (2006) supported by the presence of a trace amount of H_2 in the anode. Similarly, microbial consortia of the anode can convert CH_4 when it is diffused to the anode, thereby creating loss into the anode chamber (Rozendal et al., 2006). Despite many potential losses and unsurely of purity of gas, a membrane is mostly included in MECs.

The cathode chamber of MFCs is aerobic, whereas in MECs, it is anaerobic. The membrane present in between anode and cathode in MFCs impedes the diffusion of oxygen into the anodic chamber, whereas in MECs, this condition does not prevail. Membrane-less approaches are taking momentum in MECs as many researchers indicate that the power could be increased by over 88% from 262 mW/m^2 to 494 mW/m^2 (normalized to cathode surface area) by removing the membrane in MFCs (Liu and Logan, 2004).

The removal of the membrane also affects Coulombic efficiency (CE). CEs range from 9% to 12% in Nafion deprived MFCs, whereas with the membrane, the CEs range from 40% to 55% (Liu and Logan, 2004). The main reason behind this reduction of CE is the movement of bacteria to the cathode and the diffusion of oxygen towards the anode chamber. Out of these two issues, one issue is absent in MECs, as it operates in anaerobic conditions. Moreover, some studies indicate that high rates of production, high recoveries of hydrogen and energy are feasible using membrane-less MEC. Since H_2 is relatively not soluble in H_2O (~1.5 mg/L at 25°C and 1 bar of hydrogen partial pressure) supported by its substantial production rates, it slows down the formation of methane mediated through the microbial conversion of hydrogen (Call, 2008). Another base for opting for the membrane-less approach is that the hydrogen is insoluble in water (~1.5 mg/L at 25°C and 1 bar of hydrogen partial pressure). Moreover, if the rates of production of hydrogen are substantial, it is expected that microbially mediated translation of H_2 to CH_4 will be time-consuming (Call, 2008). These outcomes reveal that electrohydrogenesis is an efficient approach to generating H_2.

Hydrogen production was also investigated using renewable biomass in membrane-less, single-chambered MECs to avoid potential loss and to enhance the recovery of energy out of this process (Hu et al., 2008). Pure culture of *Shewanella oneidensis* MR-1 and mixed cultures were tested. These researchers suggested that by reducing the space between electrodes and escalating the ratio of electrode surface area/cell volume, the rate of volumetric hydrogen production and current density of this system increases extensively. At an applied voltage of 0.6 V in this system (single-chambered and membrane-less), hydrogen production rate with mixed culture was of 0.53 m^3/day/m^3 with a current density of 9.3 A/m^2 at pH 7 and 0.69 m^3/day/m^3 with a current density of 14 A/m^2 at pH 5.8. Methane was found in this approach, which is known to negatively affect the rate of hydrogen production. If the cathode is exposed to air, the activity of hydrogenotrophic methanogens is suppressed as methanogens that uses hydrogen as a substrate to produce methane. On the other hand, hydrogen production was stable when *S. oneidensis* was used (Hu et al., 2008).

Rozendal et al. (2006) concluded that MECs were promising and indicated that once optimized, these systems may be able to reach overall efficiencies of 90% with production rates over 10 m^3-H_2/m^3-reactor/day (Rozendal et al., 2006). There was an efficient removal of biodegradable organic matter present in wastewater employing MEC, but hydrogen recoveries were not significant (~16%). These researchers concluded that upgrading was needed for employing MECs to generate hydrogen from wastewater.

Call (2008) reported that significant recoveries of H_2 and its rates of production are feasible in a system lacking membrane. The performance of membrane-less MEC was explored in batch operation at diverse applied voltages (0.2 V < Eap < 0.8 V) employing a mixed culture and acetate as a substrate at two diverse solution conductivities (7.5 and 20 mS/cm). Overall energy recoveries employing the 7.5 mS/cm solution averaged 78 ± 4% with a maximum of 84 ± 2% at a functional voltage of 0.4 V. The competence relative to only the electrical energy input declined with applied voltage from 406 ± 6% (Eap = 0.3 V) to 194 ± 2% (Eap = 0.8 V). The utmost rate of production was 3.12 ± 0.02 m^3-H^2/m^3-reactor per day (m^3-H_2/m^3-d) at Eap = 0.8 V (7.5 mS/cm), and escalating the solution conductivity augmented the production rate for 0.3 V < Eap < 0.6 V.

7.5 FACTORS AFFECTING THE PERFORMANCE OF MICROBIAL ELECTROLYSIS CELLS

Many factors greatly affect the functioning of MECs. Prominent among them are substrate incorporated, the material involved in fabrication and operating conditions. Cathode plays a crucial role in MECs. Hence, it is pivotal that it should be cost-efficient and of higher efficiency. Platinum depicts superior performance as far as hydrogen production is considered being a cathode catalyst. However, platinum is not cost-effective that restricts its practical applicability. As a substitute to platinum, various other cost-effective catalysts have been researched that include Ni-based materials, metal nanoparticles and biocathodes. Among all these materials, Ni-based material is a better option being reasonable, substantially stable and commercially feasible. In the present scenario, the usage of MECs is not restricted to hydrogen production or wastewater treatment. However, it has expanded its utility horizons for the recovery of metals ions along with the reduction of CO_2 to formic acid (Zhao et al., 2019). There are many challenges that MECs need to overcome, which include:

- Elucidation of microbial communities involved in MEC and mechanisms of electron transfer, which are not clear yet.
- Development of highly efficient and cost-effective materials employed as electrode along with a catalyst that would boost hydrogen production.
- Curtailing the cost involved in supplying power to MEC.
- Employment of optimal sustainable clean sources of energy can be experimented like solar energy or MFCs to suffice the need of energy in MECs.

7.6 CONCLUSIONS

Hydrogen has grabbed the attention of being a clean fuel, being carbon-neutral and no adverse impact on the environment. In comparison to any other known fuel, the energy content per unit weight is highest in hydrogen. Presently, the production of biohydrogen is not cost-effective. Moreover, productivity is also not substantial. Various designs of MECs are experimented to solve this issue and the research and development are on to accomplish this target. Various substrates have also been tested that are not only cost-effective, but their utilization in MECs could help to solve the issues of wastewater treatment and water purification. In this way, hydrogen production technology proves to resolve multiple sustainability issues such as waste management, environmental reclamation and clean energy generation. However, the lack of commercially viable technologies is a major challenge that biohydrogen production has to address. Furthermore, extensive methods should be developed and organized to curtail the cost associated with the storage, transportation, and safe utilization of biohydrogen.

KEYWORDS

- **anion exchange membranes**
- **biofuels**
- **biohydrogen**
- **electrochemically active bacteria**

- **microbial dialysis cells**
- **microbial electrochemical systems**
- **microbial electrolysis cells**
- **microbial fuel cells**
- **proton exchange membrane**

REFERENCES

Balat, H., & Kirtay, E., (2010). Hydrogen from biomass—present scenario and future prospects. *Int J. Hydrogen Energy, 35*, 7416–7426.

Beer, L. L., Boyd, E. S., Peters, J. W., & Posewitz, M. C., (2009). Engineering algae for biohydrogen and biofuel production. *Curr. Opin. Biotechnol., 20*, 264–271.

Benemann, J., (1996). Hydrogen biotechnology: Progress and prospects. *Nat. Biotechnol., 14*, 1101–1103.

Bhatia, L., Sarangi, P. K., & Nanda, S., (2020). Current advancements in microbial fuel cell technologies. In: Nanda, S., Vo, D. V. N., & Sarangi, P. K., (eds.), *Biorefinery of Alternative Resources: Targeting Green Fuels and Platform Chemicals* (pp. 477–494). Springer Nature, Singapore.

Bielen, A. A. M., Verhaart, M. R. A., Van, D. O. J., & Kengen, S. W. M., (2013). Biohydrogen production by the thermophilic bacterium *Caldicellulosiruptor saccharolyticus*: Current status and perspectives. *Life, 3*, 52–85.

Call, D. F., (2008). *Hydrogen Production in a Microbial Electrolysis Cell Lacking a Membrane*. Thesis in Environmental Engineering. The Pennsylvania State University.

Cheng, S., & Logan, B. E., (2007). Sustainable and efficient biohydrogen production via electrohydrogenesis. *PNAS, 104*, 18871–18873.

Das, D., & Veziroglu, T. N., (2001). Hydrogen production by biological processes: A survey of literature. *Int J. Hydrogen Energy, 26*, 13–28.

Das, D., & Veziroglu, T. N., (2008). Advances in biological hydrogen production processes. *Int J. Hydrogen Energy, 33*, 6046–6057.

Ditzig, J., Liu, H., & Logan, B. E., (2007). Production of hydrogen from domestic wastewater using a bio-electrochemically assisted microbial reactor (BEAMR). *Int J. Hydrogen Energy, 32*, 2296–2304.

Escapa, A., San-Martín, M. I., & Morán, A., (2014). Potential use of microbial electrolysis cells in domestic wastewater treatment plants for energy recovery. *Front. Energy Res., 2*, 19.

Hallenbeck, P. C., & Ghosh, D., (2009). Advances in fermentative biohydrogen production: The way forward? *Trends Biotechnol., 27*, 287–297.

Harnisch, F., Schröder, U., & Scholz, F., (2008). The suitability of monopolar and bipolar ion-exchange membranes as separators for biological fuel cells. *Environ. Sci. Technol., 42*, 1740–1746.

Holladay, J. D., Hu, J., King, D. L., & Wang, Y., (2009). An overview of hydrogen production technologies. *Catal. Today., 139*, 244–260.

Hu, H., & Liu, Y. F., (2008). Hydrogen production using single-chamber membrane-free microbial electrolysis cells. *Water Res., 42*, 4172–4178.

Kim, J. R., Cheng, S., Oh, S. E., & Logan, B. E., (2007). Power generation using different cation, anion, and ultrafiltration membranes in microbial fuel cells. *Environ. Sci. Technol., 41*, 1004–1009.

Lee, H. S., Vermaas, W. F. J., & Rittmann, B. E., (2010). Biological hydrogen production: Prospects and challenges. *Trends Biotechnol., 28*, 262–271.

Liu, H., & Logan, B. E., (2004). Electricity generation using an air-cathode single chamber microbial fuel cell in the presence and absence of a proton exchange membrane. *Environ. Sci. Technol., 38*, 4040–4046.

Liu, H., Grot, S., & Logan, B. E., (2005). Electrochemically assisted microbial production of hydrogen from acetate. *Environ. Sci. Technol., 39*, 4317–4320.

Logan, B. E., Hamelers, B., Rozendal, R., Schroder, U., Keller, J., Freguia, S., Aelterman, P., et al., (2006). Microbial fuel cells: Methodology and technology. *Environ. Sci. Technol., 40*, 5181–5192.

Min, B., Kim, J., Oh, S., Regan, J. M., & Logan, B. E., (2005). Electricity generation from swine wastewater using microbial fuel cells. *Water Res., 39*, 4961–4968.

Nanda, S., & Berruti, F., (2021a). A technical review of bioenergy and resource recovery from municipal solid waste. *J. Hazard. Mater., 403*, 123970.

Nanda, S., & Berruti, F., (2021b). Municipal solid waste management and landfilling technologies: A review. *Environ. Chem. Lett., 19*, 1433–1456.

Nanda, S., & Berruti, F., (2021c). Thermochemical conversion of plastic waste to fuels: A review. *Environ. Chem. Lett., 19*, 123–148.

Nanda, S., Azargohar, R., Dalai, A. K., & Kozinski, J. A., (2015). An assessment on the sustainability of lignocellulosic biomass for biorefining. *Renew Sust. Energ. Rev., 50*, 925–941.

Nanda, S., Dalai, A. K., Gökalp, I., & Kozinski, J. A., (2016). Valorization of horse manure through catalytic supercritical water gasification. *Waste Manag., 52*, 147–158.

Nanda, S., Mohammad, J., Reddy, S. N., Kozinski, J. A., & Dalai, A. K., (2014). Pathways of lignocellulosic biomass conversion to renewable fuels. *Biomass Conv. Bioref., 4*, 157–191.

Nanda, S., Rana, R., Hunter, H. N., Fang, Z., Dalai, A. K., & Kozinski, J. A., (2019). Hydrothermal catalytic processing of waste cooking oil for hydrogen-rich syngas production. *Chem. Eng. Sci., 195*, 935–945.

Nanda, S., Rana, R., Zheng, Y., Kozinski, J. A., & Dalai, A. K., (2017). Insights on pathways for hydrogen generation from ethanol. *Sustain. Energy Fuels, 1*, 1232–1245.

Nandi, R., & Sengupta, S., (1998). Microbial production of hydrogen: An overview. *Crit. Rev. Microbiol., 24*, 61–84.

Nath, K., & Das, D., (2004). Improvement of fermentative hydrogen production: Various approaches. *Appl Microbiol. Biotechnol., 65*, 520–529.

Okolie, J. A., Nanda, S., Dalai, A. K., Berruti, F., & Kozinski, J. A., (2020). A review on subcritical and supercritical water gasification of biogenic, polymeric and petroleum wastes to hydrogen-rich synthesis gas. *Renew. Sust. Energy Rev., 119*, 109546.

Okolie, J. A., Rana, R., Nanda, S., Dalai, A. K., & Kozinski, J. A., (2019). Supercritical water gasification of biomass: A state-of-the-art review of process parameters, reaction mechanisms and catalysis. *Sustain. Energy Fuels, 3*, 578–598.

Rahimnejad, M., Adhami, A., Darvari, S., Zirepour, A., & Oh, S. E., (2015). Microbial fuel cell as new technology for bioelectricity generation: A review. *Alexandria Eng. J., 54*, 745–756.

Reddy, S. N., Nanda, S., Vo, D. V. N., Nguyen, T. D., Nguyen, V. H., Abdullah, B., & Nguyen-Tri, P., (2020). Hydrogen: Fuel of the near future. In: Nanda, S., Vo, D. V. N., & Nguyen-Tri, P., (eds.), *New Dimensions in Production and Utilization of Hydrogen* (pp. 1–20). Elsevier, Cambridge, USA.

Redwood, M. D., Paterson-Beedle, M., & MacAskie, L. E., (2009). Integrating dark and light bio-hydrogen production strategies: Towards the hydrogen economy. *Rev. Environ. Sci. Biotechnol., 8*, 149–185.

Rivera, I., & Patil, S. A., (2019). Microbial electrolysis for biohydrogen production: Technical aspects and scale up experiences. In: Mohan, S. V., Varjani, S., & Pandey, A., (eds.), *Microbial Electrochemical Technology* (pp. 871–898). Elsevier.

Rozendal, R. A., Hamelers, H. V. M., Euverink, G. J. W., Metz, S. J., & Buisman, C. J. N., (2006). Principle and perspectives of hydrogen production through biocatalyzed electrolysis. *Int. J. Hydrogen Energy, 31*, 1632–1640.

Rozendal, R. A., Hamelers, H. V. M., Molenkamp, R. J., & Buisman, C. J. N., (2007). Performance of single chamber biocatalyzed electrolysis with different types of ion-exchange membranes. *Water Res., 41*, 1984–1994.

Rozendal, R. A., Jeremiasse, A. W., Hamelers, H. V. M., & Buisman, C. J. N., (2008). Hydrogen production with a microbial biocathode. *Environ. Sci. Technol., 42*, 629–634.

Rupprecht, J., Hankamer, B., Mussgnug, J. H., Ananyev, G., Dismukes, C., & Kruse, O., (2006). Perspectives and advances of biological H_2 production in microorganisms. *Appl. Microbiol. Biotechnol., 72*, 442–449.

Sarangi, P. K., & Nanda, S., (2018). Recent developments and challenges of acetone-butanol-ethanol fermentation. In: Sarangi, P. K., Nanda, S., & Mohanty, P., (eds.), *Recent Advancements in Biofuels and Bioenergy Utilization* (pp. 111–123). Springer Nature, Singapore.

Sarangi, P. K., & Nanda, S., (2019). Recent advances in consolidated bioprocessing for microbe-assisted biofuel production. In: Nanda, S., Sarangi, P. K., & Vo, D. V. N., (eds.), *Fuel Processing and Energy Utilization* (pp. 141–157). CRC Press, Boca Raton.

Sarangi, P. K., & Nanda, S., (2020). Biohydrogen production through dark fermentation. *Chem. Eng. Technol., 43*, 601–612.

Sarangi, P. K., Nanda, S., & Vo, D. V. N., (2020). Technological advancements in the production and application of biomethanol. In: Nanda, S., Vo, D. V. N., & Sarangi, P. K., (eds.), *Biorefinery of Alternative Resources: Targeting Green Fuels and Platform Chemicals* (pp. 127–140). Springer Nature.

Sarangi, P. K., Singh, T. A., & Singh, N. J., (2017). Agricultural crop residues: Unused biomass having huge energy potential. In: Ghosal M., (ed.), *Contemporary Renewable Energy Technologies for Sustainable Agriculture* (pp. 47–61). Narosa Publishing House, New Delhi, India.

Sarangi, P. K., Singh, T. A., & Singh, N. J., (2019). Pineapple as potential crop resource: Perspective and value addition. In: Singh, T. A., Sarangi, P. K., & Sarangthem, N., (eds.), *Food Bioresources and Ethnic Foods of Manipur* (pp. 83–91). North-east, India. Empyreal Publishing House; India.

Schubert, C., (2006). Microbiology: Batteries not included circuits of slime. *Nature, 441*, 277–279.

Selembo, P. A., Perez, J. M., Lloyd, W. A., & Logan, B. E., (2009). High hydrogen production from glycerol or glucose by electrohydrogenesis using microbial electrolysis cells. *Int. J. Hydrogen Energy, 34*, 5373–5381.

Singh, S., Kumar, R., Setiabudi, H. D., Nanda, S., & Vo, D. V. N., (2018). Advanced synthesis strategies of mesoporous SBA-15 supported catalysts for catalytic reforming applications: A state-of-the-art review. *Appl. Catal. A: Gen., 559*, 57–74.

Winfield, J., Gajda, I., Greenman, J., & Ieroulos, I., (2016). A review into the use of ceramics in microbial fuel cells. *Bioresour. Technology, 215*, 296–303.

Zhao, F., Harnisch, F., Schroder, U., Scholz, F., Bogdanoff, P., & Herrmann, I., (2006). Challenges and constraints of using oxygen cathodes in microbial fuel cells. *Environ. Sci. Technol., 40*, 5193–5199.

Zhao, W., & Ci, S., (2019). Nanomaterial as electrode materials of microbial electrolysis cell for hydrogen generation. *Micro Nano Technol.*, 213–242.

Zuo, Y., Cheng, S., Call, D., & Logan, B. E., (2007). Tubular membrane cathodes for scalable power generation in microbial fuel cells. *Environ. Sci. Technol., 41*, 3347–3353.

Zuo, Y., Maness, P. C., & Logan, B. E., (2006). Electricity production from steam-exploded corn stover biomass. *Energy Fuels, 20*, 1716–1721.

CHAPTER 8

Confluence of Nanocatalysts and Bioenergy: An Overview of Microbial Electrochemical Systems and Biohydrogen Production

PIYUSH PARKHEY, KUSH NAYAK, REECHA SAHU, and ARUNIMA SUR

Amity Institute of Biotechnology, Amity University Chhattisgarh, Raipur, Chhattisgarh, India
E-mail: pparkhey@rpr.amity.edu (Piyush Parkhey)

ABSTRACT

Energy sources from biological sources such as microbial electrochemical systems and biohydrogen have received considerable attention in recent times of ever-increasing fuel crisis and environmental pollution. Microbial electrochemical systems have huge potential in a myriad of industrial, environmental, and biomedical applications. Several optimization studies have been undertaken to enhance the yield and improve their productivity. Even though desirable results have been obtained. However, to attain an industrial level performance, the efficiency needs to be increased significantly. Improvement of the electrode architecture is, therefore, quite substantial to enhance the efficacy of the microbial electrochemical systems. Biohydrogen production, on the other hand, though economic, suffers from low yield and productivity. The use of nanoparticles as a catalyst to enhance biohydrogen productivity appears to be a feasible solution. This chapter discusses the electron transfer mechanism in electrochemically active bacteria and summarizes the studies conducted to identify novel electrode construction materials and assemblies for microbial electrochemical systems. This is followed by a detailed discussion on the use of nanoparticles as efficient catalysts and their mechanism in improving biohydrogen evolution from fermentative reactions.

8.1 INTRODUCTION

Bioenergy, in simple terms, refers to the renewable form of energy that is derived from biological sources rich in organic matter such as energy crops, biomass, agro-wastes, and recently algae and seaweeds. The rising concerns of running out of fossil fuels, environmental issues associated with greenhouse gas emissions have mandated exploration of alternative sources of energy that are sustainable, environmentally safe, and readily available. Biomass-based fuels have been a frontrunner in this exploration, and significant advancements in this domain have been reported in the past two decades.

Of the various biomass-based biofuels, noteworthy developments have been described in the bioenergy produced from microorganisms via conversion of chemical energy to electrical energy in microbial fuel cells (MFCs). These MFCs have since been modified to operate under different conformations to realize multiple other usages a common term for all such systems has been defined as Microbial electrochemical systems (MESs). They, owing to their multi-faceted applications have garnered significant scientific attention in recent times. Few major applications, other than bioelectricity for which MESs have been used are for the production of biohydrogen or biomethane (microbial electrolytic cells, MEC), synthesis of industrially important compounds by CO_2 reduction (microbial electro synthetic cells MESCs) and *in situ* desalination of water (MDCs). Other significant applications of microbial electrochemical systems include biosensing of pollutants, contaminants, and toxins, wastewater treatment, lab-on-a-chip point of care diagnostic devices, microsized medical implants, etc. These applications are based on the common principle of oxidation of substrates by the exoelectrogenic bacteria in the anodic chamber, subsequent generation and flow of electrons to the cathodic chamber. This principle is used with slight modifications and in the different architecture of microbial electrochemical systems for the number of applications that these systems are known for.

Several factors regulate the performance of microbial electrochemical systems. Considering the multitude of applications of MESs in energy, environmental, medical, and other allied sectors, various studies have been undertaken to optimize these factors to obtain maximum performance output. These works report modifying different parameters such as electrode features, electrochemical cell architecture, working conditions, optimal microbial cultures and mode of cell operation. A detailed comparative analysis of the various structural and process parameters of a microbial electrolytic cell for hydrogen production has been earlier reviewed (Parkhey and Gupta, 2017), and it is believed that these parameters regulate the performance of other microbial electrochemical systems in more or less the same way.

The thermodynamic principles operational within any microbial electrochemical system limit the rate of electron transfer and in turn, control the maximum performance output. Therefore, even with continual process optimizations, the maximum practical yield from the system remains much less than the theoretical maximum. The major factors accountable for this are ohmic and activation losses within the system. The ohmic losses due to electrodes correspond to the resistance of the movement of electrons through the biofilm to the anode, electrical connections, and the conducting wires. Similarly, the activation losses occur due to the energy lost in establishing electric current across the reactor while overcoming the energy barrier across the electrode/electrolyte interface. While reducing electrode distances and using a high concentration of electrolytes are suggested for decreasing the ohmic losses, the overpotential losses can be overcome by modifying the electrode surface, which can yield higher current density. Therefore, taking into account the fact that of the different factors that regulate the output from any microbial electrochemical system, the electrode geometry, material, and architecture play a major role in maximizing the performance, many studies have been reported that optimize MES system as a function of electrode performance. The most prominent of these modifications have been coating the electrode surface with catalysts to enhance the rate of electron transfer, thereby reducing the action overpotentials. However, surface modifications of electrodes with catalysts significantly increases the overall input costs of the process.

Alternatively, nanomodification of electrodes has been proposed as a cost-effective, efficient, and sustainable approach to further enhance the output of microbial electrochemical systems. In this chapter, we have discussed the various nanocatalysts that have been developed to modify the electrode surface to facilitate better electron transport and increase the efficiency of the system. A discussion on the mechanism of the electron transfer is presented which would help understand the rationale behind various nanomodifications in the electrode. Finally, a discussion on the development of nanocatalysts to enhance biohydrogen production is also presented.

8.2 MECHANISM OF ELECTRON EXCHANGE

The ability of certain anaerobic microorganisms to transfer electrons across long distances to an electrode as a part of their respiration and establish an electric current has been exploited for various practical purposes. Detailed knowledge of this ability, along with the added knowledge of the transport kinetics of charge and electrode chemistry between the interface of cell and electrode has been critical in developing existing applications as well as in developing new applications. Such microorganisms, which are also commonly, called electrochemically active bacteria (EAB) or exoelectrogens serve as a microbial electrocatalyst, and the process of electron transport is termed as extracellular electron transfer (EET). Several mechanisms have been proposed that describe the process of EET in different exoelectrogenic bacteria. Most of these mechanisms have been elucidated in *Geobacter* species, which predominantly colonize the anodic electrode or *Shewanella* species which are easier to lab cultivate and facultative aerobes, i.e., can use oxygen as electron acceptor (Lovley and Nevin, 2011).

The different mechanisms of EET can be classified into direct and indirect EET. Direct EET corresponds to the electron transfer which occurs via cytochromes present in the outer cell membrane or redox enzymes contained within them when the cell membrane and the electrodes are in direct contact. Another mechanism of direct EET is via conductive pili-like structures, which are also known as microbial nanowires (Gorby et al., 2006; Kumar et al., 2017; Reguera et al., 2005). The indirect EET involves the secretion of compounds such as flavins, quinones, and phenazines by exoelectrogenic bacteria as redox-active molecules that transfer electrons to the electrodes. The electrochemical analyzes and gene expression studies reveal that in *Geobacter* spp. the electron conduction in the electroactive biofilms over the electrodes occurs through the conductive pili (type IV) and *c*-type cytochrome and not via electron shuttles. Initially, it was believed that the type IV pili in *Geobacter* species are composed of PilA protein. However, a recent study by Wang et al. (Wang et al., 2019) revealed that the nanowires are assembled by polymerization of hexaheme cytochrome OmcS, which elucidates the noteworthy capacity of the *Geobacter* species for long-distance electron transfer.

In *Shewanella* spp., such conductive pili which can transfer electrons directly from the inner membrane to the cell exterior, do not play any significant role in extracellular electron transfer. Therefore, no direct contact is required. The electron conduction occurs through the electron shuttles such as quinones or flavins, which upon reduction from cytochromes are released from the cell and carry electrons to the electrode. However, surface nanowires in *Shewanella* sp. play a critical role in electron transfer by acting as a p-type semiconductor as revealed in studies using conducting-probe atomic force microscopy (CP-AFM) and field-effect transistors (FETs) (Leung et al., 2013).

Strains of *Pseudomonas* sp. use excreted mediators such as phenazines for electron transfer, while other naturally occurring compounds such as humic acids have also been reported to be effective mediators (Kumar et al., 2017). Figure 8.1 describes the different models for the suggested mechanisms for electron transfer to electrodes. In studies with other non-exoelectrogenic bacteria, the presence of mediators such as methylene blue was reported to be effective in electron transport (Sugnaux et al., 2013). The detailed mechanism of cytochromes, conducive pili, nanowires electron shuttles have been delineated in detail by Lovley and Nevin (2011) and Kumar et al. (2017). The details are not discussed here as it would be a mere replication and because it is beyond the scope of this chapter's domain. However, interested readers are suggested to read the related reviews for more insights.

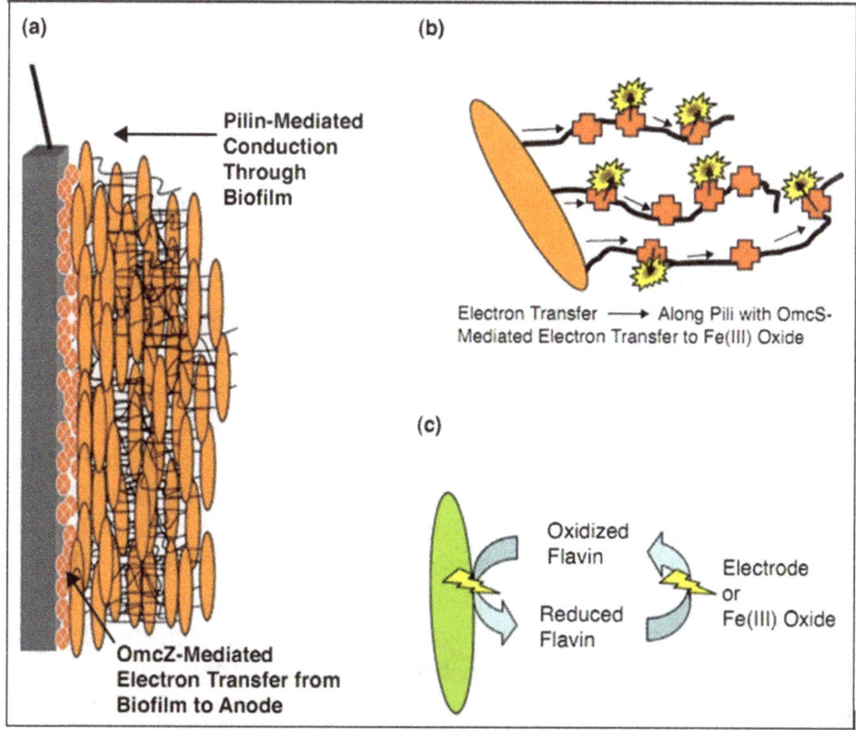

FIGURE 8.1 Suggested models for the mechanism of electron transfer from exoelectrogenic bacteria to the electrode surface.

Source: Reproduced with permission from Lovley and Nevin (2011). © Elsevier.

8.3 NANOELECTRODE DEVELOPMENT IN MICROBIAL ELECTROLYTIC CELLS

With the mechanism of electron transfer in the microorganism-electrode interface became well known, added by the knowledge of cell attachment to other conductive materials, further advancement studies concentrated on improving the electron transfer rate and enhance the microbial electrochemical system's efficiency. The development of nanoelectrodes for microbial electrochemical systems have been suggested as a practical solution towards the quest of improving the system's output. Nanomaterials, other than providing a remarkably large surface-area-to-volume ratio also offer unique electrochemical properties for better electron transfer to the electrode (Erbay, 2016). In the following section, we discuss the important developments in the MES performance that have been reported in the past decade as a function of nanomodification of the electrode.

The first report of its kind was presented by Jiang et al. (2010) who studied the mechanism of transfer of electrons in *Shewanella oneidensis* MR-1 using an optically transparent nanoelectrode-based platform. High-resolution imaging revealed that initially the current is not correlated to the number of cells while a mediated electron transfer occurs post biofilm formation. This study opened a new avenue for the study of microorganism-electrode interaction for paved the way for future research in developing better electrode materials.

A comparative analysis of different combinations of nanostructures to expand the surface area of the electrode was reported by Alatraktchi et al. (2012), in which the gold and titanium nanoparticles were deposited as nanograss over Silicium and carbon paper electrodes. A power output of 0.002 mW/m^2 was obtained with plain Silicium electrodes which elevated to 2.5 mW/m^2 by coating Ti/Au nanoparticles as nanograss on one side and further to 86.0 mW/m^2 on coating both the sides. On depositing Au-nanoparticles on carbon-paper electrodes the reported power density was 346.9 mW/m^2, almost 3-times that of plain carbon-paper (Al Alatraktchi et al., 2012). A similar 3-D bimetallic Pd-Cu nanodendrites was studied as an Oxygen reduction reaction (ORR) catalyst (Xiong et al., 2013). However, the electrode was not used in a microbial electrochemical system but the electrochemical studies revealed superior performance in oxygen reduction reaction, which can make it an efficient, economic, and stable catalyst in MFCs.

The utilization of nanoelectrodes in MESs for effective wastewater remediation as a means for environmental remediation purposes has also been proposed. A novel Au@TiO_2 nanocomposite was used by Kalathil et al. (2013) for green and sustainable azo dye degradation. This system degraded the dye more effectively than with plain electrodes. Toxicity tests also showed that the treated wastewater was almost non-toxic with nanocomposite-assisted electrodes. The efficient and quick mineralization further confirms the effectiveness of nanomaterials in increasing the efficacy of MES output. Usage of nanoelectrodes made of platinum and titanium wires have also been assessed for simultaneous synthetic wastewater treatment with power production in an MFC (Torabiyan et al., 2014). The total chemical oxygen demand (COD) removal with nano Pt-electrodes and Ti-nanowires was 98.3% as compared to 72.5% for plain electrodes and wires. Denitrification of wastewater is another important domain of environmental restoration. A report by Wang et al. (2016) described the use of a novel titanium cathode for denitrification of groundwater which resulted in a 154% higher nitrate reduction than the untreated electrode.

Jiang et al. (2014) reported the application of biogenic inorganic nanoparticles to enhance the EET in MFCs. They reported the use of iron sulfide for nano-decoration of the Ti/Au electrode by the process of biomineralization and subsequent cell/FeS nanoparticle aggregation of the bacterial cell. Their results showed that crystalline FeS nanoparticles formed a uniform coating over the cell membrane, which resulted in improved electron transfer at the cell/electrode interface, thereby increasing the current output.

There have also been attempts that did not yield satisfactory or desirable results. Tan (2014) reported the use of gold nanowires to increase the electrode's available surface area and thereby improving electrical conductivity. However, the surface of electrode post-modification did not significantly increase the biofilm formation of *Shewanella oneidensis* MR-1. Thus, no remarkable improvement in MFC output was observed. The probable reason for this could be due to the formation of pockets of unavailable space and air by nanostructures that were not used by the bacteria.

Other than environmental analysis, MESs have also been used as biosensor tools to detect pathogens and as the micro-power source for implantable medical devices. One such study to directly determine *Streptomyces* spp. by electrochemical detection using a nanomaterial-based microbial sensor was described by Hasan et al. (2014). They reported the use of multi-walled carbon nanotubes at a concentration of 10% w/w to yield the highest electrocatalytic performance with a good correlation between the population of metabolically active *Streptomyces* spp. and the oxidation peak current.

In a slightly different approach, Sedki et al. (2017) for the first time reported a nanocomposite made of reduced graphene oxide-hyperbranched chitosan (rGO-HBCs) to detect *Pseudomonas aeruginosa* as a pathogen. Their study demonstrated detection of the pathogen at a limit that was 10 times lower than the previous reports when cell viability was tested at varying concentrations of antibiotics such as Kanamycin, Simvastatin, and Ciprofloxacin. This and various other studies have confirmed the superior performance of MES using graphene and graphene-modified electrocatalysts such as GO and graphene-based CNTs, nanosheets, and composites (Shaari and Kamarudin, 2017).

In another study, a gold nanoelectrode array was used in an enzymatic glucose biofuel cell which could provide micro-watts power to implantable bioelectronic devices (Kulkarni et al., 2015). The sensor device made with pyramidal Au nanotip of the effective surface area of 0.04 cm^2 yielded a maximal power output of 112.21 $\mu W/cm^2$ using 20 mM glucose as a substrate. These studies confirm that MES performance for sensing as well as micro-power production improves significantly with the use of nanoelectrodes.

A noteworthy and unique study to develop synthetic biological nanowires was conducted by Tan et al. (2016). In this work, the microbial nanowires of *Geobacter sulfurreducens* were genetically engineered to incorporate more *trp* residues to facilitate faster electron transport. Their work resulted in synthetic nanowires where half to 1.5 nm yielding a 2000 times increase in conductivity reduced the diameter. This study demonstrated that electrically conductive nanowires can be biologically synthesized which can significantly improve the conduction capabilities of the pili.

Remarkable progress in the designing and development of nanoelectrodes has taken place in the last couple of years, where interdisciplinary research has shown meaningful results. For example, harvesting photosynthetic electrons from plants and algae by the insertion of nanoelectrodes directly into the living algal cells has been proposed to be a highly efficient and environmentally friendly approach for the generation of renewable energy (Hong et al., 2018). The mentioned work discusses the fabrication of an Au/ Si_3N_4 deposited cantilever nanoelectrode by modifying a commercial AFM tip. The nanoelectrodes upon insertion into the algal cells displayed a photosynthetic current density of 6 mA/cm^2, which is multiple times higher than the presently used methods of isolated thylakoid membranes. Therefore, this study is quite significant concerning its findings regarding a sustainable and environmentally friendly approach to energy harvesting and promises replication of this approach in plant or algal-based MES systems.

In another innovative work, the researchers describe the development of a hierarchically porous nitrogen-doped CNTs/rGO composite as an MFC anode (Wu et al., 2018). While the 3D porous structure provided ample surface area for bacterial growth, the nitrogen doping allowed improved flavin redox reactions at the electrode surface. The maximum power density with the doped composite as the anodic electrode was 8.9 times higher than the carbon cloth and higher than CNTs, N-rGO, and undoped CNT/rGO. The study confirms that both biofilm formation and better electron transfer increases the MES output. These outcomes further provide insights to develop better electrode systems for practical electrochemical applications.

In another study, the stability of carbon nanotubes at the anode and their effect on the electrochemical properties and anodic biofilms in MFCs was analyzed (Zhang et al., 2018). The study reported the preparation of a novel CNT-modified graphite felt (CNTs-GF) electrode, which showed a stable performance for over a year with a sustainable power production of 2±0.1 W/m^2. The long terms stable power production confirms that CNTs-GF supported the growth of electrochemically active biofilms and enhanced the electron transfer rate. Therefore, electrode modification using carbon nanotubes could be an interesting avenue for exploring better electrode architecture with sustained long-term MES output.

Considering the importance of nanomaterials in electrode fabrication and decoration, it becomes imperative to determine their characteristics to optimize their properties to obtain a better electrocatalyst. Wilde et al. (2018) reported a novel technique for investigating the electrocatalytic properties of nanoparticles. They used nickel-iron layered double hydroxide (NiFe LDH) nanoparticles etched in carbon nanoelectrodes. The catalyst could be immobilized below the picograms level and offered a high resolution for the faradic current response. Cyclic voltammetry was used to analyze the intrinsic activity of the nanocatalyst. Their reported method could be used to determine the stability and perceive the catalytic properties of the catalyst.

With the advancements in the electrode development technologies, different techniques and applications of MES have come up. In one such novel study, Zhang et al. (2019) described the preparation of composite made-up of polypyrrole/chitosan/carbon nanotubes for the capacitive deionization for removing copper ions. They obtained a specific capacitance of 103.2 F/g and an adsorption capacity of 16.8 mg/g, which was multiple times higher than the otherwise composite electrode.

An altogether different direction to the possibilities of developing better electrodes in designing better microbial electrochemical devices was depicted by the findings of Freyman et al. (2019). In their work, *Shewanella oneidensis* MR-1 live cells were combined in ink, which was used for 3D printing of anode. The living bacterial anode demonstrated significant results in azo dye degradation. This recent report unlocks newer areas of research and possibilities for designing better electrode materials and electrochemical cells. A different, yet unique method for azo dye degradation was proposed by Zhang et al. (2020). They developed a Pt/TiO_2/graphene/polyethylene sheet by a facile thermal curing process and reported 100% degradation of acetaminophen after a treatment of 50 min while maintaining a 90% removal rate even after recycling the electrode five times.

With the intent of increasing the efficiency of MFC by improving the flavin-based interfacial electron transfer, Wang et al. (2020) described a hierarchical porous carbon (CPC) derived from an aerogel from a mixture of cellulose, NaOH, and urea. Their results demonstrated enhanced power generation, attributable to the enhanced surface area and double layer capacitance due to the micropores generated during the reaction of urea and carbon. Not just cellulose, but also various other polymeric materials that have been reported to be intrinsically conductive have been used for nanoelectrode preparation. One such conducting polymer is polyaniline, which owing to its high stability, sustained electrical conductivity and variable protonation and oxidation states have been reported by many researchers as a competitive candidate for nanoelectrode development (Kausar, 2020).

A rather new study was conducted that focussed on improving the limited charge transfer between the whole cell-electrode surface. The study conducted by Reggente et al. (2020) described the electrosynthesis of poly(3,4-ethylene dioxythiophene) (PEDOT)-based electrodes in the presence of SDS dopant. This electrode successfully captured photosynthetically derived electrons under multiple cycles of dark and photoperiods by the phototrophic bacteria *Synechocystis* sp. PCC 6803. This study opened up new perspectives for designing enhanced microbial-based devices using such non-conventional electrodes, which can catalyze direct as well as non-direct electron transfer.

The development of novel electrodes has not just been limited toward the anodic end. Various studies have reported superior electrochemical performance as a function of better cathodic electrodes. In one such study, Priya et al. (2020) used MnO_2@rGO as a cathode electrode with different Antimony-Tin combinations on rGO to study MFC performance. Another report was from Zhang et al. (2020), who developed and used a Fe_2O_3/NiO photocathode in a methanol fuel cell and reported higher photoelectric conversion. They postulated that the photogenerated electrons are accumulated on the modified electrode surface and induce the ORR. The photocatalytic methanol fuel cell reported better performance under conditions of illumination. This study opened up new horizons in the area of solar-chemical energy conversions in a fuel cell. A recent study described the development of activated carbon catalysts doped with nitrogen, embedded with iron-based clusters using the ball milling method. The electrochemical analysis showed higher ORR for these catalysts with a maximum power output of 2387 mW/m^2, higher than reported in any study so far (Zhang et al., 2020).

8.4 NANOCATALYST AND BIOHYDROGEN

In the previous sections, we have discussed how nanocatalyst has been developed to enhance the performance of microbial electrochemical systems for bioenergy production in the form of bioelectricity and other applications such as biosensing and bio-electrosynthesis. Over the past one and a half decades, biohydrogen has emerged as another interesting alternative to conventional fuels. Over the next sections, we have discussed the importance and evolution of biohydrogen as a bioenergy source and how the development of nanocatalysts has improved its production via various routes.

8.4.1 BIOHYDROGEN

Hydrogen, owing to its attributes of clean combustion profile, high-energy-density and ease of operation at ambient temperatures has already been considered as the most ideal replacement for fossil fuels (Parkhey and Gupta, 2017). However, the current commercial hydrogen processes such as catalytic steam reforming, methane oxidation, and coal gasification have significantly high-energy demands and again are dependent largely on existing fossil fuels. Therefore, they do not exactly solve the very problem for which the solution is being envisaged. Biohydrogen or biologically produced hydrogen fixes these drawbacks, as its production is cost as well as energy-efficient. It corresponds to the hydrogen produced via biological routes by a microorganism such as bacteria, archaebacteria, cyanobacteria, or eukaryotic organisms such as algae using biomass and other waste materials as the substrate. While bacteria (and archaebacteria) produce bacteria by dark fermentative routes, cyanobacteria, and algae produce hydrogen by photo fermentation and bio-photolysis of water, respectively. Although energy-efficient than industrial processes, due to thermodynamic limitations, these methods of biohydrogen production are of no match for the other biological fuels, which can be produced from the same biomass. This seriously limits the commercial acceptability of these methods. Over the years, multiple optimization studies have been conducted to improve biohydrogen production in terms of yield and rate, using fermentative approaches to match up the industrial demands. The use of nanocatalysts has been one such technique for simpler, efficient, and economic biohydrogen production.

8.4.2 ROLE OF NANOCATALYSTS AND ASSOCIATED MECHANISMS

As it became evident that fermentative hydrogen production has thermodynamic limitations to maximum yield, alternative approaches to enhance the biological hydrogen production were explored. The use of metal and metal oxide nanoparticles (NPs) has been advocated by many researchers to be an ideal practice to obtain higher biohydrogen yield. These NPs are proposed to increase hydrogen evolution in biological reactions by the following:

- Increasing the intracellular electron transfer.
- Supplementing the H_2 evolving hydrogenase enzymes.
- Selectively promoting the growth of hydrogen producers in a mixed microflora.

However, since their biocompatibility is questionable, many researchers have reported and prescribed the use of biologically synthesized nanoparticles as a more favorable approach (Patel et al., 2018). The application of both organic as well as inorganic nanoparticles has been proposed for enhancing biohydrogen production. A comparative analysis of hydrogen production rate and yield by various organic and inorganics nanoparticles has already been tabulated by Patel et al. (2018) and Shanmugam et al. (2020a). Here, we have summarized the role and possible mechanism by which nanoparticles improve hydrogen production. A tabulated summary of such a proposed mechanism is presented in Table 8.1. A diagrammatic representation of the mechanism by which nanoparticles enhance the hydrogen production by dark or photo-fermentation is depicted in Figure 8.2, as described by Shanmugam et al. (2020a).

TABLE 8.1 Proposed Mechanism of Various Nanoparticles and Their Respective Roles in Enhancing Hydrogen Production via Fermentative Routes

Biohydrogen production route	Class	Type	Mechanism	References
Dark fermentation	Inorganic	Iron/Iron oxide (Fe/ Fe_2O_3 NPs)	Increase the activity of the Fe-Fe or Ni-Fe hydrogenase enzyme	Akia et al. (2014); Lin et al. (2016); Taherdanak et al. (2016)
		Zero valent iron oxide (Fe^0) NPs	Enrichment of microbial community with biohydrogen producers and promote ferredoxin and hydrogenase activity	Yang and Wang (2018)
		Nickel and Nickel-graphene (Ni/Ni-C NPs)	Enhancement of Ni-Fe hydrogenase activity and growth of relevant bacterial species	Elreedy et al. (2017)
		Silver (Ag NPs)	Redirection of carbon flux from ethanol to acetic acid pathway and reduction in the lag phase of hydrogen evolution	Zhao et al. (2013)
		Gold (Au NPs)	Alteration of metabolite production (acetate > butyrate and low ethanol)	Zhang and Shen (2007)
Photo-fermentation	Inorganic	Titanium oxide (TiO_2 NPs)	Protein and polysaccharide decomposed to small organic compounds, photosynthetic bacterial growth promoted, nitrogenase activity improved H_2-uptake hydrogenase activity diminished.	Zhao and Chen (2011)
		Iron (Fe NPs)	Increase in photosynthetic pigments resulting in increased microalgal biomass	Eroglu et al. (2013)

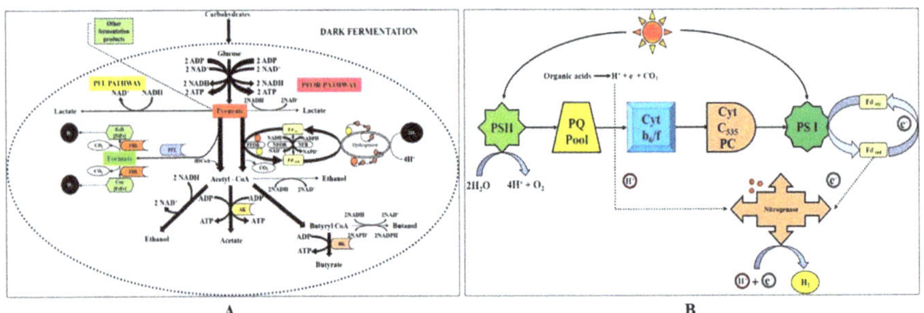

FIGURE 8.2 Proposed mechanism of improvement of biohydrogen production using nanoparticles via (A) dark fermentation and (B) photo-fermentation.

Source: Reproduced with permission from Shanmugam et al. (2020a). © Elsevier.

8.5 NANOCATALYTIC ENHANCEMENT OF BIOHYDROGEN EVOLUTION

The first report on studying the effect of nanocatalysts on biohydrogen production from biomass was put forward by Chang et al. (2011) wherein an electrophoretic deposition method was used to design a Pt-Ru bimetallic nanocatalyst and applied as a functionalized mesoporous media on carbon textile. Using an aqueous phase reforming reactor, the catalysts were used for biomass to biohydrogen conversion. The efficient and economic catalysts yielded hydrogen at a maximal rate of 0.47 mmol/g of cellulose. Even though the production rate was maximum at a high temperature of 235°C. The method still appeared a feasible one-pot process for the conversion of low-cost agro-waste substrate to hydrogen.

Another novel approach to study nano-catalyzed hydrogen production was put forward by Khan et al. (2013). They reported the generation of electrons and protons by sodium acetate decomposition using an electrochemically active biofilm. The electrons generated were used for the reduction of Au^{3+} to Au^0 for nanoparticle formation, which in turn were used to catalyze the proton reduction by electrons produced earlier. They reported hydrogen production at a maximal rate of ~105±2 mL/L reactor volume/day in a novel, economic, and environmentally safe manner.

Unconventional support for NPs was proposed by Wimonsong et al. (2014) wherein Au nanocatalysts supported on Au/Fe-Zn-Mg-Al-O hydrotalcite were synthesized, and their effect on the improvement of fermentative hydrogen production was studied. Their results demonstrated that Au/Zn-Mg-Al HT yielded hydrogen at 2.74±0.14 mol H_2/mol sucrose. The higher production can be ascribed to the release of gold on zinc surface, acting as the active site for the major fermentative hydrogen production catalyst, i.e., hydrogenase

A similar study was conducted by Le and Nitisoravut (2015) by preparing Ni-Mg-Al nanocatalyst on hydrotalcite support by the co-precipitation method. A dosage of 250 mg/L Ni-Mg-Al HT increased the yield of hydrogen by 80% resulting in the production of 3.37±0.17 mol H_2/mol sucrose. This study further confirmed that the release of metals on the surface was decisive in improving the hydrogen production, as in this case, Ni formed an important elemental component of Ni-Fe hydrogenase while Mg, other than being an important component of cell wall and membrane is known to stimulate many electron transfer reactions thereby increasing hydrogen evolution.

Metal oxide nanoparticles can also enhance fermentative hydrogen production by strengthening specific metabolic pathways while weakening others. This was aptly demonstrated by Lin et al. (2016) using ferric oxide nanoparticles. They suggested that FONPs enhanced the acetate pathway in *Enterobacter aerogenes* while weakening the ethanol pathway, which allowed more of nicotinamide adenine dinucleotide (NAD) to be redirected for reducing proton to hydrogen thereby increasing the yield from 164.5±2.3 mL/g to 192.4±1.1 mL/g glucose.

The activity of nanocatalysts for improving hydrogen production has not just been reported with simpler substrates or pure compounds but also with complex organic substrates such as wastewaters. Gadhe et al. (2015) studied how hydrogen evolution by dark fermentation using complex distillery wastewater as a substrate is affected by nickel oxide (NiO) and hematite (Fe_2O_3) nanoparticles. Their results revealed that co-addition of the two NPs enhanced the maximum yield (8.8 mmol/g COD) and hydrogen production rate (18.1 mmol/g VSS.d) by about 62% and 221% respectively. This increase can be credited to the improved activity of ferredoxin oxidoreductase and hydrogenase at the interface of iron oxide and Nickle oxide interface.

Elreedy et al. (2017) studied hydrogen production from mono-ethylene glycol polluted industrial wastewater and the effect of Ni NPs and Ni-graphene nanocomposites (Ni-Gr NC) using a mixed bacterial culture. At an optimal concentration of 60 mg/L for Ni NPs and Ni-Gr NC hydrogen production improved by a substantial 23% and 105% respectively. However, at a higher dosage of NPs or NC, hydrogen production decreased indicating the potentially toxic effects of the heavy metal. They also conducted a separate study to evaluate the effects of NiO, ZnO, and Fe_2O_3 NPs either individually or in combination on hydrogen evolution by mixed culture bacteria isolated from sludge of a municipal wastewater treatment plant (Elreedy et al., 2019). Their results depict that NPs in combination was most effective in enhancing the hydrogen evolution rate to a maximum of 13±0.8 mL/L/h with the highest volumetric production of hydrogen of 150±9 mL/L. Significantly lower concentrations of acetaldehyde and ethanol with comparatively higher acetic acid suggest that ethanol was converted to acetaldehyde and subsequently to acetic acid with associated production of hydrogen. Another noteworthy inference of the study was an increase in hydrogen-evolving *Clostridiales* in the fermentation reaction carried out in the presence of nanoparticles.

A unique approach was proposed by Kodhaiyoli et al. (2019) in which *Boerhavia diffusa* mediated green synthesis approach was described for bimetallic Co-Ni nanoparticles. The fabricated Co-Ni NPs could successfully catalyze biohydrogen evolution by *Citrobacter freundii* on glucose substrate yielding a net biohydrogen of 0.26 mol H_2/mol glucose. In another one of its kind approach, Shanmugam et al. (2020b) used Fe_3O_4@SiO_2-chitosan nanoparticles as support for the immobilization of fungal laccase for biohydrogen production from delignified lignocellulosic biomass.

Many specific reports have also been proposed about optimizing the photo-fermentation reactions using nanocatalysts. Pandey et al. (2015) prescribed the use of TiO_2 NPs for photo-fermentative hydrogen production using *Rhodobacter sphaeroides* NMBL-02. Their findings revealed that in the presence of pyruvate as a carbon source, a maximum H_2 evolution of 1900 mL/L was obtained. Using the same photo-bacteria, in a separate report, $Fe(SO_4)OH(H_2O)_2$ NPs were used to report a 1.2-fold increment in H_2 evolution to a maximum of 2046 mL H_2/L of medium (Dolly et al., 2015). Nanocatalyst enhancement of hydrogen production has also been proposed with microalgal culture.

Giannelli and Torzilo (2012) reported the use of microalga *Chlamydomonas reinhardtii* in a photobioreactor added with the suspension of silica NPs, which assisted in uniform light distribution to the cells and reported a mean H_2 yield of 3121.5 ± 178.9 mL. Other than the work discussed earlier, there have been several other studies that have successfully reported the use of nanoparticles for enhancing biohydrogen production either by dark or photo fermentation.

Nanocatalysts have not only been used for the improvement of hydrogen production but also biogas, specifically methane production. One such report was presented by Anjum et al. (2018) wherein nanocatalysts which were developed from visible light active ZnO-ZnS/Ag_2O-Ag_2S, mediated solubilization of sludge was done with a reported improvement in methane production by 54% with cumulative production increasing to 0.6985 mL/g VS and reduced start-up time for methanogenesis from 23 to 13 days. A comparative analysis of these published reports is presented in Table 8.2.

TABLE 8.2 Comparative Analysis of Different Nanoparticles and Their Effect on Biohydrogen Production and Yield

Nanoparticle	Process features	Biohydrogen production rate and yield	Remarks	Reference
Photothermal biomaterial designated GeO_2-SiO_2-Chitosan-Medium-LaB_6	Photo-fermentation carried out by a photosynthetic bacterial (PSB) biofilm	• 2.9 mmol/h/m^2	• The biomaterial was designed to increase the overall utilization of the light by LaB_6 and transmitted light was optimum for PSB growth	Li et al. (2017)
Hematite nanoparticle	Fermentative H_2 production from sucrose wastewater in a CSTR	• The maximum rate of H_2 production was 5.9 L/L/d	• Acid incubation enhanced the granulation process which led to an increase in H_2 evolution	Salem et al. (2017)
Maghemite nanoparticles coated with sludge	Sequential dark-photo fermentation from starch wastewater and pre-treated sludge without NPs was control	• Overall H_2 yield: 166.8±27.8 mL H_2/g COD removed	• Reactor with sludge immobilized on NPs showed enhanced H_2 yield as compared to control	Nasr et al. (2015)
Low-cost ferrihydrite nanorods	Dark fermentative H_2 production by *Clostridium pasteurianum* using glucose as a substrate with magnetite and hematite NP as control	• The maximum cumulative H_2 production was 1.03 mmol (68.9% higher). • The maximum yield was 3.55 mol H_2/mol glucose (15.6% higher)	• Ferrihydrite was suggested to enhance glucose conversion efficiency and hydrogenase activity.	Zhang et al. (2019)
TiO_2 and Fe nanoparticles at different concentrations	Dark fermentation using *Clostridium pasteurianum* on xylose	• 24.9% increase in H_2 production rate (8.7 H_2-L/L/d) at 50 ppm Fe NP.	• No significant increase of overall H_2 yield by Fe NPs suggesting the NPs did not affect the enzyme activity.	Hsieh et al. (2016)

TABLE 8.2 *(Continued)*

Nanoparticle	Process features	Biohydrogen production rate and yield	Remarks	Reference
Fe^0 and Ni^0 Nanoparticles against Fe^{2+} and Ni^{2+} ions	Mesophilic dark H_2 fermentation by anaerobic sludge pretreated by heat shock	• Ni^{2+}, Fe^0 NPs and Fe^{2+} increased H_2 yield by 55%, 37% and 15% respectively. • Ni^0 effect was insignificant	• Soluble metabolite analysis revealed ethanol and propionate production reduced H_2 yield while butyrate formation favored the process	Taherdanak et al. (2016)
Mesoporous Fe_3O_4 NPs	Activation of biohydrogen production capacity from an alkaline shock treated mixed culture	• Maximum cumulative H_2 production was 26% higher than control with 400 mg/L Fe_3O_4 NPs	• Decreased H_2 production lag phase with mesoporous Fe_3O_4 nanoparticles to 12 h (50 h less than control)	Zhao et al. (2011)
Fe (II) oxide and nickel oxide NPs	Batch culture. Glucose-fed anaerobic thermophilic dark fermentation	• Fe(II) oxide and nickel oxide NPs increased H_2 yield by 34.38% and 5.47%, respectively.	• Metabolite analysis revealed H_2 production as acetic acid pathway-related.	Engliman et al. (2017)
Photocatalytic TiO_2, ZnO, and SiC nanoparticles	Photo fermentation by *Rhodopseudomonas* sp. nov. strain A7 on acetate substrate	• H_2 yield (TiO_2 NP): 2.8 mol H_2/mol acetate. • H_2 yield (ZnO NP): 2.6 mol H_2/mol acetate • H_2 yield (SiC NP): 2.99 mol H_2/mol acetate • Maximum volume: 2272 mL H_2/L culture	• Sic NP showed the better potential of stimulating photo fermentative hydrogen evolution as compared to TiO_2 and ZnO NP	Liu et al. (2017)

TABLE 8.2 *(Continued)*

Nanoparticle	Process features	Biohydrogen production rate and yield	Remarks	Reference
Nickel NP (optimal concentration 5.67 mg/L	Mesophilic fermentation using glucose as a substrate	• Maximum cumulative biohydrogen production: 4400 mL. • Biohydrogen yield of 2.5 mol of H_2/mol of glucose	• Process optimization carried out by response surface methodology with central composite designs	Mullai et al. (2013)
Green synthesized Fe NPs using *Murraya koenigii* leaf extract as a reducing and stabilizing agent	Dark fermentative H_2 production using *Clostridium acetobutylicum* NCIM 2337 from glucose	• Maximum H_2 yield: 2.3±0.09 mol H_2/mol glucose • Total H_2 content: 52±0.8% • Maximum production rate: 25.3 mL/h	• FeNPs are suggestive of increasing the ferredoxin activity leading to higher hydrogen production.	Mohanraj et al. (2014)
Fe^0 and Ni^0 NPs	Dark fermentation at mesophilic temperature using starch as substrate and anaerobic sludge pretreated with heat-shock as inoculum	• Maximum H_2 yield: 149.8 mL/g VS (200% higher than non-catalyzed control) at Fe^0 and Ni^0, i.e., 37.5 mg/L	• Process optimized by response surface methodology. • Metabolite analysis suggested hydrogen evolution through the acetic acid pathway.	Taherdanak et al. (2015)

8.6 CONCLUSIONS

The availability of energy resources is a key parameter that measures the socio-economic advancements of a nation. The ever-increasing energy demands, supplemented by the apprehensions of exhaustion of fossil fuels and environmental concerns of greenhouse emissions have delegated for the exploration, development, and employment of alternative energy sources. The chapter discussed microbial electrochemical systems and biohydrogen as two alternative bioenergy sources that can replace the conventional sources of energy. Microbial electrochemical systems are a potential alternative towards the production of clean, renewable, and sustainable bioenergy, synthesis of industrially important compounds, environmental remediation, pathogen sensing and toxin detection. However, the maximum output performance of the systems is limited by the rate of electron transfer from microorganisms to the electrode surface. Improving the effective surface area, increasing the electrical conductivity of the electrode material, enhancing the biocompatibility of the electrode surface for better biofilm formation would minimize many glitches associated with MES performance efficiency. An additional factor regarding the commercialization of MES could be the availability of the material for electrode fabrication.

Superior performing cathode catalysts such as nano-decorated electrode surfaces and identifying better analysis techniques for investigating electrode performance is therefore utmost required to realize the wider application of microbial electrochemical cells. The prospects include identifying novel and better materials with higher electron transport efficiency, developing new materials, and exploring better analytical techniques to determine the efficacy of MES. Extensive, focused, and interdisciplinary research and development spanning the biological sciences, material science, nanoscience, and physical science should be conducted to obtain practical and applicable results. The production of biohydrogen has come a long way since it was first reported as an economic and efficient energy source. However, many glitches persist in their production and storage that limit their commercial utilization. Nanoparticle catalyzed biohydrogen production can be a way of enhancing the net yield and production of biohydrogen systems. However, to compete against other commercial methods of hydrogen production, biohydrogen production processes still need significant improvements.

KEYWORDS

- **chemical oxygen demand**
- **conducting-probe atomic force microscopy**
- **electrochemically active bacteria**
- **extracellular electron transfer**
- **field-effect transistors**
- **microbial electrochemical systems**
- **microbial fuel cells**
- **nanoelectrodes**
- **oxygen reduction reaction**

REFERENCES

Akia, M., Yazdani, F., Motaee, E., Han, D., & Arandiyan, H., (2014). A review on conversion of biomass to biofuel by nanocatalysts. *Biofuel Res. J., 1*, 16–25.

Al Alatraktchi, F. A. Z., Zhang, Y., Noori, J. S., & Angelidaki, I., (2012). Surface area expansion of electrodes with grass-like nanostructures and gold nanoparticles to enhance electricity generation in microbial fuel cells. *Bioresour. Technol., 123*, 177–183.

Anjum, M., Kumar, R., Al-Talhi, H. A., Mohamed, S. A., & Barakat, M. A., (2018). Valorization of biogas production through disintegration of waste activated sludge using visible light ZnO-ZnS/Ag_2O-Ag_2S photocatalyst. *Process Saf. Environ. Prot., 119*, 330–339.

Chang, A. C. C., Louh, R. F., Wong, D., Tseng, J., & Lee, Y. S., (2011). Hydrogen production by aqueous-phase biomass reforming over carbon textile supported Pt-Ru bimetallic catalysts. *Int. J. Hydrogen Energy, 36*, 8794–8799.

Dolly, S., Pandey, A., Pandey, B. K., & Gopal, R., (2015). Process parameter optimization and enhancement of photo-biohydrogen production by mixed culture of *Rhodobacter sphaeroides* NMBL-02 and *Escherichia coli* NMBL-04 using Fe-nanoparticle. *Int. J. Hydrogen Energy, 40*, 16010–16020.

Elreedy, A., Fujii, M., Koyama, M., Nakasaki, K., & Tawfik, A., (2019). Enhanced fermentative hydrogen production from industrial wastewater using mixed culture bacteria incorporated with iron, nickel, and zinc-based nanoparticles. *Water Res., 151*, 349–361.

Elreedy, A., Ibrahim, E., Hassan, N., El-Dissouky, A., Fujii, M., Yoshimura, C., & Tawfik, A., (2017). Nickel-graphene nanocomposite as a novel supplement for enhancement of biohydrogen production from industrial wastewater containing mono-ethylene glycol. *Energy Convers. Manag., 140*, 133–144.

Engliman, N. S., Abdul, P. M., Wu, S. Y., & Jahim, J. M., (2017). Influence of iron (III) oxide nanoparticle on biohydrogen production in thermophilic mixed fermentation. *Int. J. Hydrogen Energy, 42*, 27482–27493.

Erbay, C., (2016). *Micro/nano Technologies for Achieving Sustainable Microbial Electrochemical Cell Systems.* Texas A&M University.

Eroglu, E., Eggers, P. K., Winslade, M., Smith, S. M., & Raston, C. L., (2013). Enhanced accumulation of microalgal pigments using metal nanoparticle solutions as light filtering devices. *Green Chem., 15*, 3155.

Freyman, M. C., Kou, T., Wang, S., & Li, Y., (2019). 3D printing of living bacteria electrode. *Nano Res., 12*, 1–5.

Gadhe, A., Sonawane, S. S., & Varma, M. N., (2015). Influence of nickel and hematite nanoparticle powder on the production of biohydrogen from complex distillery wastewater in batch fermentation. *Int. J. Hydrogen Energy, 40*, 10734–10743.

Giannelli, L., & Torzillo, G., (2012). Hydrogen production with the microalga *Chlamydomonas reinhardtii* grown in a compact tubular photobioreactor immersed in a scattering light nanoparticle suspension. *Int. J. Hydrogen Energy, 37*, 16951–16961.

Gorby, Y. A., Yanina, S., McLean, J. S., Rosso, K. M., Moyles, D., Dohnalkova, A., Beveridge, T. J., et al., (2006). Electrically conductive bacterial nanowires produced by *Shewanella oneidensis* strain MR-1 and other microorganisms. *PNAS, 103*, 11358–11363.

Hassan, R. Y. A., Hassan, H. N. A., Abdel-Aziz, M. S., & Khaled, E., (2014). Nanomaterials-based microbial sensor for direct electrochemical detection of *Streptomyces* spp. Sensors *Actuators, B Chem., 203*, 848–853.

Hong, H., Kim, Y. J., Han, M., Yoo, G., Song, H. W., Chae, Y., Pyun, J. C., Grossman, A. R., & Ryu, W. H., (2018). Prolonged and highly efficient intracellular extraction of photosynthetic electrons from single algal cells by optimized nanoelectrode insertion. *Nano Res., 11*, 397–409.

Hsieh, P. H., Lai, Y. C., Chen, K. Y., & Hung, C. H., (2016). Explore the possible effect of TiO_2 and magnetic hematite nanoparticle addition on biohydrogen production by *Clostridium pasteurianum* based on gene expression measurements. *Int. J. Hydrogen Energy, 41*, 21685–21691.

Jiang, X., Hu, J., Fitzgerald, L. A., Biffinger, J. C., Xie, P., Ringeisen, B. R., & Lieber, C. M., (2010). Probing electron transfer mechanisms in *Shewanella oneidensis* MR-1 using a nanoelectrode platform and single-cell imaging. *PNAS, 107*, 16806–16810.

Jiang, X., Hu, J., Lieber, A. M., Jackan, C. S., Biffinger, J. C., Fitzgerald, L. A., Ringeisen, B. R., & Lieber, C. M., (2014). Nanoparticle facilitated extracellular electron transfer in microbial fuel cells. *Nano Lett., 14*, 6737–6742.

Kalathil, S., Lee, J., & Cho, M. H., (2013). Catalytic role of Au@TiO_2 nanocomposite on enhanced degradation of an azo-dye by electrochemically active biofilms: A quantized charging effect. *J. Nanoparticle Res., 15*(1), 1–6.

Kausar, A., (2020). High-performance competence of polyaniline-based nanomaterials. *Mater. Res. Innov., 24*, 113–122.

Khan, M. M., Lee, J., & Cho, M. H., (2013). Electrochemically active biofilm mediated bio-hydrogen production catalyzed by positively charged gold nanoparticles. *Int. J. Hydrogen Energy, 38*, 5243–5250.

Kodhaiyolii, S., Mohanraj, S., Rengasamy, M., & Pugalenthi, V., (2019). Phytofabrication of bimetallic Co-Ni nanoparticles using *Boerhavia diffusa* leaf extract: Analysis of phytocompounds and application for simultaneous production of biohydrogen and bioethanol. *Mater. Res. Express, 6*, 095051.

Kulkarni, T., Gupta, D., & Slaughter, G., (2015). An enzymatic glucose biofuel cell based on Au nano-electrode array. *IEEE Sensors - Proc.*, 5–8.

Kumar, A., Hsu, L. H. H., Kavanagh, P., Barrière, F., Lens, P. N. L., Lapinsonnière, L., Lienhard, J. H., et al., (2017). The ins and outs of microorganism-electrode electron transfer reactions. *Nat. Rev. Chem., 1*, 1–13.

Le, D. T. H., & Nitisoravut, R., (2015). Ni-Mg-Al hydrotalcite for improvement of dark fermentative hydrogen production. *Energy Proc., 79*, 301–306.

Leung, K. M., Wanger, G., El-Naggar, M. Y., Gorby, Y., Southam, G., Lau, W. M., & Yang, J., (2013). *Shewanella oneidensis* MR-1 bacterial nanowires exhibit p-type, tunable electronic behavior. *Nano Lett., 13*, 2407–2411.

Li, Y., Zhong, N., Liao, Q., Fu, Q., Huang, Y., Zhu, X., & Li, Q., (2017). A biomaterial doped with LaB_6 nanoparticles as photothermal media for enhancing biofilm growth and hydrogen production in photosynthetic bacteria. *Int. J. Hydrogen Energy, 42*, 5793–5803.

Lin, R., Cheng, J., Ding, L., Song, W., Liu, M., Zhou, J., & Cen, K., (2016). Enhanced dark hydrogen fermentation by addition of ferric oxide nanoparticles using *Enterobacter aerogenes*. *Bioresour. Technol., 207*, 213–219.

Liu, B., Jin, Y., Wang, Z., Xing, D., Ma, C., Ding, J., & Ren, N., (2017). Enhanced photo-fermentative hydrogen production of *Rhodopseudomonas* sp. nov. strain A7 by the addition of TiO_2, ZnO and SiC nanoparticles. *Int. J. Hydrogen Energy, 42*, 18279–18287.

Lovley, D. R., & Nevin, K. P., (2011). A shift in the current: New applications and concepts for microbe-electrode electron exchange. *Curr. Opin. Biotechnol., 22*, 441–448.

Mohanraj, S., Kodhaiyolii, S., Rengasamy, M., & Pugalenthi, V., (2014). Green synthesized iron oxide nanoparticles effect on fermentative hydrogen production by *Clostridium acetobutylicum*. *Appl. Biochem. Biotechnol., 173*, 318–331.

Mullai, P., Yogeswari, M. K., & Sridevi, K., (2013). Optimisation and enhancement of biohydrogen production using nickel nanoparticles - a novel approach. *Bioresour. Technol., 141*, 212–219.

Nasr, M., Tawfik, A., Ookawara, S., Suzuki, M., Kumari, S., & Bux, F., (2015). Continuous biohydrogen production from starch wastewater via sequential dark-photo fermentation with emphasize on maghemite nanoparticles. *J. Ind. Eng. Chem., 21*, 500–506.

Pandey, A., Gupta, K., & Pandey, A., (2015). Effect of nanosized TiO_2 on photofermentation by *Rhodobacter sphaeroides* NMBL-02. *Biomass and Bioenergy, 72*, 273–279.

Parkhey, P., & Gupta, P., (2017). Improvisations in structural features of microbial electrolytic cell and process parameters of electrohydrogenesis for efficient biohydrogen production: A review. *Renew. Sust. Energy Rev., 69*.

Patel, S. K. S., Lee, J. K., & Kalia, V. C., (2018). Nanoparticles in biological hydrogen production: An overview. *Indian J. Microbiol., 58*, 8–18.

Priya, A., Deva, S., Shalini, P., & Pydi, S. Y., (2020). Antimony-tin based intermetallics supported on reduced graphene oxide as anode and MnO_2@rGO as cathode electrode for the study of microbial fuel cell performance. *Renew. Energy, 150*, 156–166.

Reggente, M., Politi, S., Antonucci, A., Tamburri, E., & Boghossian, A. A., (2020). Design of optimized PEDOT-based electrodes for enhancing performance of living photovoltaics based on phototropic bacteria. *Adv. Mater. Technol., 5*, 1900931.

Reguera, G., McCarthy, K. D., Mehta, T., Nicoll, J. S., Tuominen, M. T., & Lovley, D. R., (2005). Extracellular electron transfer via microbial nanowires. *Nature, 435*, 1098–1101.

Salem, A. H., Mietzel, T., Brunstermann, R., & Widmann, R., (2017). Effect of cell immobilization, hematite nanoparticles and formation of hydrogen-producing granules on biohydrogen production from sucrose wastewater. *Int. J. Hydrogen Energy, 42*, 25225–25233.

Sedki, M., Hassan, R. Y. A., Hefnawy, A., & El-Sherbiny, I. M., (2017). Sensing of bacterial cell viability using nanostructured bioelectrochemical system: rGO-hyperbranched chitosan nanocomposite as a novel microbial sensor platform. *Sensors Actuators, B Chem., 252*, 191–200.

Shaari, N., & Kamarudin, S. K., (2017). Graphene in electrocatalyst and proton conducting membrane in fuel cell applications: An overview. *Renew. Sust. Energy Rev., 69*, 862–870.

Shanmugam, S., Hari, A., Pandey, A., Mathimani, T., Felix, L. O., & Pugazhendhi, A., (2020a). Comprehensive review on the application of inorganic and organic nanoparticles for enhancing biohydrogen production. *Fuel, 270*, 117453.

Shanmugam, S., Krishnaswamy, S., Chandrababu, R., Veerabagu, U., Pugazhendhi, A., & Mathimani, T., (2020b). Optimal immobilization of *Trichoderma asperellum* laccase on polymer-coated Fe_3O_4@SiO_2 nanoparticles for enhanced biohydrogen production from delignified lignocellulosic biomass. *Fuel, 273*, 117777.

Sugnaux, M., Mermoud, S., Da Costa, A. F., Happe, M., & Fischer, F., (2013). Probing electron transfer with *Escherichia coli*: A method to examine exoelectronics in microbial fuel cell type systems. *Bioresour. Technol., 148*, 567–573.

Taherdanak, M., Zilouei, H., & Karimi, K., (2015). Investigating the effects of iron and nickel nanoparticles on dark hydrogen fermentation from starch using central composite design. *Int. J. Hydrogen Energy, 40*, 12956–12963.

Taherdanak, M., Zilouei, H., & Karimi, K., (2016). The effects of Fe^0 and Ni^0 nanoparticles versus Fe^{2+} and Ni^{2+} ions on dark hydrogen fermentation. *Int. J. Hydrogen Energy, 41*, 167–173.

Tan, D. S., (2014). Modification of a graphite electrode surface with nanomaterial for use in microbial fuel cells. *Fluid Mech. Conf. AFMC.*, 19–22.

Tan, Y., Adhikari, R. Y., Malvankar, N. S., Pi, S., Ward, J. E., Woodard, T. L., Nevin, K. P., et al., (2016). Synthetic biological protein nanowires with high conductivity. *Small, 12*, 4481–4485.

Torabiyan, A., Nabi, B. G. R., Mehrdadi, N., & Javadi, K., (2014). Application of nano-electrode platinum (Pt) and nano-wire titanium (Ti) for increasing electrical energy generation in microbial fuel cells of synthetic wastewater with carbon source (Acetate). *Int. J. Environ. Res., 8*, 453–460.

Wang, D., Wang, Y., Yang, J., He, X., Wang, R. J., Lu, Z. S., & Qiao, Y., (2010). Cellulose aerogel derived hierarchical porous carbon for enhancing flavin-based interfacial electron transfer in microbial fuel cells. *Polymers, 12*, 664.

Wang, F., Gu, Y., O'Brien, J. P., Yi, S. M., Yalcin, S. E., Srikanth, V., Shen, C., Vu, D., Ing, N.L., Hochbaum, A.I., Egelman, E.H., & Malvankar, N.S., (2019). Structure of microbial nanowires reveals stacked hemes that transport electrons over micrometers. *Cell, 177*, 361–369.e10.

Wang, L., Li, M., Feng, C., Hu, W., Ding, G., Chen, N., & Liu, X., (2016). Ti nanoelectrode fabrication for electrochemical denitrification using box-Behnken design. *J. Electroanal. Chem., 773*, 13–21.

Wilde, P., Barwe, S., Andronescu, C., Schuhmann, W., & Ventosa, E., (2018). High resolution, binder-free investigation of the intrinsic activity of immobilized NiFe LDH nanoparticles on etched carbon nanoelectrodes. *Nano Res., 11*, 6034–6044.

Wimonsong, P., Nitisoravut, R., & Llorca, J., (2014). Application of Fe-Zn-Mg-Al-O hydrotalcites supported Au as active nano-catalyst for fermentative hydrogen production. *Chem. Eng. J., 253*, 148–154.

Wu, X., Qiao, Y., Shi, Z., Tang, W., & Li, C. M., (2018). Hierarchically porous N-doped carbon nanotubes/reduced graphene oxide composite for promoting flavin-based interfacial electron transfer in microbial fuel cells. *ACS Appl. Mater. Interfaces, 10*, 11671–11677.

Xiong, L., Huang, Y. X., Liu, X. W., Sheng, G. P., Li, W. W., & Yu, H. Q., (2013). Three-dimensional bimetallic Pd-Cu nanodendrites with superior electrochemical performance for oxygen reduction reaction. *Electrochim. Acta, 89*, 24–28.

Yang, G., & Wang, J., (2018). Improving mechanisms of biohydrogen production from grass using zero-valent iron nanoparticles. *Bioresour. Technol., 266*, 413–420.

Zhang, D., Wang, Y., Wang, Y., Zhang, Y., & Song, X. M., (2020). Fe_2O_3/NiO photocathode for photocatalytic methanol fuel cell: An insight on solar energy conversion. *J. Alloys Compd., 815*, 152377.

Zhang, Q., Yao, C., Hong, J. M., & Chang, C. T., (2020). Preparation of Pt/TiO_2/graphene/polyethylene sheets via a facile molding process for azo dye electrodegradation. *J. Nanosci. Nanotechnol., 20*, 3287–3294.

Zhang, X., Guo, X., Wang, Q., Zhang, R., Xu, T., Liang, P., & Huang, X., (2020). Iron-based clusters embedded in nitrogen-doped activated carbon catalysts with superior cathodic activity in microbial fuel cells. *J. Mater. Chem. A., 8*, 10772–10778.

Zhang, Y. J., Xue, J. Q., Li, F., Dai, J. Z., & Zhang, X. Z. Y., (2019). Preparation of polypyrrole/chitosan/carbon nanotube composite nano-electrode and application to capacitive deionization process for removing Cu^{2+}. *Chem. Eng. Process. - Process Intensif., 139*, 121–129.

Zhang, Y., & Shen, J., (2007). Enhancement effect of gold nanoparticles on biohydrogen production from artificial wastewater. *Int. J. Hydrogen Energy, 32*, 17–23.

Zhang, Y., Chen, X., Yuan, Y., Lu, X., Yang, Z., Wang, Y., & Sun, J., (2018). Long-term effect of carbon nanotubes on electrochemical properties and microbial community of electrochemically active biofilms in microbial fuel cells. *Int. J. Hydrogen Energy, 43*, 16240–16247.

Zhang, Y., Xiao, L., Wang, S., & Liu, F., (2019). Stimulation of ferrihydrite nanorods on fermentative hydrogen production by *Clostridium pasteurianum*. *Bioresour. Technol., 283*, 308–315.

Zhao, W., Zhang, Y., Du, B., Wei, D., Wei, Q., & Zhao, Y., (2013). Enhancement effect of silver nanoparticles on fermentative biohydrogen production using mixed bacteria. *Bioresour. Technol., 142*, 240–245.

Zhao, W., Zhao, J., Chen, G. D., Feng, R., Yang, J., Zhao, Y. F., Wei, Q., Du, B., & Zhang, Y.F., (2011). Anaerobic biohydrogen production by the mixed culture with mesoporous Fe_3O_4 nanoparticles activation. *Adv. Mater. Res., 306, 307*, 1528–1531.

Zhao, Y., & Chen, Y., (2011). Nano-TiO_2 Enhanced photofermentative hydrogen produced from the dark fermentation liquid of waste activated sludge. *Environ. Sci. Technol., 45*, 8589–8595.

Index

A

Acetaminophen, 198
Acetic acid, 16, 56, 60, 79, 80, 91, 98, 101, 134, 204
Acetyl coenzyme A (acetyl-CoA), 134, 136, 152
Acidogenic bacteria, 131
Active fermentation, 149
Adenosine
  diphosphate, 133
  triphosphate (ATP), 132, 133, 136, 147, 148, 150, 158
Adsorption capacity, 59, 61, 197
Aerobic respiration, 133
Aerogel, 51
Aerosols, 13, 40
Agglomeration, 44, 58, 112
Agricultural waste, 34
Agro-residues, 177
Agro-waste, 7, 202
Aldehyde, 2, 39
Alkali catalysts, 47
Alkaline electrolyzer, 143
Allothermal, 7, 8, 40, 42, 45, 46
  gasification, 7, 8
Ammonia, 2, 78, 154, 181
Anabaena
  cylindrica, 130
  variabilis, 130, 154
Anaerobes, 151
Anaerobic
  bacteria, 134
  conditions, 149, 179, 182
  digestion, 23
  fermentation, 158, 166
  microorganisms, 22, 133, 134, 191
Anion exchange membranes (AEMs), 181, 184
Anode acidification, 181
Anoxygenic photosynthetic reactions, 147
Anthropogenic wastes, 176
Aromatic rings, 17
Asphaltene, 10
Autothermal, 7, 8, 16, 17, 33, 40, 42, 55, 78, 91, 92, 102
  catalytic mode, 16
  gasification, 7, 8
  operating conditions, 33
  reactors, 8
  reforming (ATR), 17, 54, 55, 78, 91, 92, 102
  steam reforming (ATSR), 17
  updraft gasification, 40
Azo dye degradation, 195, 198

B

Bacterial chlorophyll-a (BChl-a), 150, 151
Bacteriopheophytin, 150
Ball milling method, 199
Benzene, 40
Bi-functional catalyst, 60
Biocatalysis, 161, 166
Biochar, 14
Biochemical, 35, 128
  conversion rate, 20
Biodiesel, 91, 166
Bioelectricity, 190, 199
Bio-electrochemical reactions, 130
Bio-electrosynthesis, 199
Bioenergy, 189, 190, 203
Biofuels, 26, 110, 128, 136, 142, 184, 190
Biohydrogen, 1–9, 11–26, 35, 102, 109, 127–136, 141, 146, 168, 169, 177, 178, 184, 189–191, 199, 200, 202, 204, 205, 208
  evolution, 189, 202, 204
  opportunities and challenges, 25
  production, 1, 3, 4, 6–9, 11–23, 25, 26, 127, 130–132, 135, 168, 177, 178, 184, 191, 200, 202, 204, 208
    autothermal catalytic transformation, 16
    bio-oil reforming, 19

catalyst-based steam transformation, 15
electro-chemical catalytic transformation, 18
fast pyrolysis, 19
fermentation technology, 23
gasification, 6
hydrothermal gasification, 9
light-dependent biosynthetic pathway, 21
light-independent biosynthetic pathway, 22
mono-step production, 17
plasma mediated biomass gasification, 13
pyrolysis, 14
sequential cracking, 18
solar gasification, 12
steam assisted biomass gasification, 11
Biohythane, 135
Biomass, 3–15, 17, 18, 20, 23–25, 33–37, 39–56, 63, 64, 78, 80, 97–99, 109–112, 114–123, 127–129, 131, 133, 135, 136, 143, 145, 146, 158, 162, 164, 166, 176, 177, 182, 190, 199, 200, 202, 204
gasification, 7, 8, 12, 14, 42, 47
catalysts, 47
characteristics, 42
equivalence ratio, 46
pressure, 45
steam-to-biomass ratio, 45
temperature, 43
moisture content, 63
pyrolysis, 14, 20, 36
Biomedical applications, 189
Biomethane, 133, 190
Biomineralization, 195
Bio-oil, 4, 5, 14–20, 54, 56, 62, 98, 111, 114
reforming, 18
transportation, 20
Biophotolysis, 21, 130, 132–134
Bioreactor, 23, 132, 134, 135, 175
Biosensing, 190, 199
Bio-waste, 143, 168
Boudouard, 7, 40, 43, 44, 47, 115, 120
reaction, 7, 40, 43, 115
Bubbling, 15, 41–43, 45, 52, 53
Butyric acid, 134

C

Carbohydrates, 26, 128, 146, 147, 158
Carbon, 1–3, 6, 9–12, 16–18, 20, 22, 25, 26, 34, 36, 40, 43, 49–53, 56, 57, 62, 78, 91, 92, 101, 102, 109, 111, 112, 120, 122, 128, 132, 141, 142, 144, 145, 152, 160–163, 165–169, 176, 177, 184, 194, 195, 197–199, 202, 204
conversion, 18, 43, 50–53, 120
decomposition, 57
dioxide emission, 33–35, 55, 57, 63, 64, 146
emissions, 1, 2, 128, 142, 168, 169, 176
monoxide (CO), 12, 26
methanation reaction, 7
neutral
process, 141
renewable resource, 3
substrate, 165
Carbonaceous
compounds, 26
materials, 113
Carboxylic acid, 20
Carcinogenic compounds, 50
Carotenoids, 130
Catalysts, 8–10, 12, 18, 20, 35, 36, 47–52, 55, 56, 60, 63, 64, 79, 80, 87, 91, 93, 96, 98, 101, 109, 121, 122, 145, 183, 189, 191, 199, 202, 208
to-sorbent ratio, 33
Catalytic
action, 134
composition, 110
conversion, 77, 102, 111
reactions, 15
steam reforming, 79
Cathode, 98, 130, 143, 178–183, 195, 198, 208
alkalization, 181
Cathodic electrodes, 198
Cationic exchange membrane (CEM), 181
Cell
growth, 22, 167
membrane, 192, 195
operation, 190
Cellular flux, 151
Cellulose, 9, 10, 14, 17, 26, 42, 43, 111, 114, 133, 162, 198, 202
Cellulosic biomass, 133
Channeling, 113

Char
combustion, 12, 42
generation, 9
Chemical
looping, 19, 33, 54, 57, 61, 78, 94, 96, 97, 101, 102
combustion (CLC), 19
gasification (CLG), 52–54
reforming (CLR), 19, 54, 55, 57, 62, 78, 94, 96, 97, 101, 102
steam reforming (CLSR), 19, 57, 61–63
water splitting (CLWS), 58
oxygen demand (COD), 195, 203, 209
Chlorophyll, 130, 150
Circulating fluidized bed (CFB), 8, 61
Citrobacter freundii, 204
Clostridium, 134, 151, 152, 158, 159, 162, 163, 168
articum, 134
CnHm decomposition, 57
CNT-modified graphite felt (CNTs-GF), 197
Co-gasification, 109, 115–123
Combined reforming of methane (CMR), 78, 102
Commercialization, 5, 52, 156, 175, 208
Conducting-probe atomic force microscopy, 209
Conventional
hydrocarbon fuels, 2, 128
pyrolysis, 20
resources, 34
sources, 34, 35, 208
Copper-cesium catalyst, 80
Co-precipitation method, 49, 60, 203
Corrosion, 40
Coulombic efficiency (CE), 182
Cracking reaction, 86
Critical temperature, 10
Crystallinity, 129
Cyanobacteria, 130, 144, 146–150, 153–155, 157–160, 169, 199
Cyanobacterial cellular system, 153
Cyclic
electron transport, 150
redox mechanism, 33
voltammetry, 197
Cyclone, 53
Cytochrome, 147, 150, 192, 193

D

Dark
condition, 151
fermentation, 22, 23, 127, 129–136, 141, 142, 144, 146, 148, 151–153, 158, 160, 161, 165, 178, 202, 203
fermenting bacteria, 160
Decoupled dual loop gasification system, 12
Department of biotechnology (DBT), 169
Desulfurization, 2
De-volatilization, 11, 113, 114
Dielectric barrier discharge, 13
Disproportionation reaction, 86
Dry
biomass (DB), 53
reforming, 54, 55, 78, 86–88, 91–93, 101, 102
of methane (DMR), 78, 86, 87, 92, 102
Dual fluidized bed reactor (DFBR), 42, 48, 49, 64

E

Effluents, 51, 132, 141, 142, 168
Elector acceptor, 146
Electric field, 145
Electrical conductivity, 195, 198, 208
Electroactive biofilms, 192
Electrochemical
analysis, 199
catalytic reforming (ECR), 18
cell, 198, 208
architecture, 190
process, 143, 145
systems, 35, 135, 189
Electrochemically active bacteria (EAB), 180, 184, 189, 192, 209
Electrode, 13, 182, 191–195, 198
interface, 194, 195
surface, 182, 191, 193, 197–199, 208
Electrogenic bacteria, 130, 131
Electrohydrogenesis, 178, 179, 182
Electrolysis, 2, 14, 35, 78, 98, 129–131, 136, 143, 145, 146, 175, 176, 178–180, 185
Electrolytes, 191
Electrolyzer, 143, 145
Electromagnetic radiation, 98

Electron, 18, 21, 22, 130, 132, 144, 146, 147, 149–151, 155, 157–160, 169, 178–180, 184, 189–200, 202, 203, 208, 209
  acceptor, 21, 147, 192
    molecule, 21
  exchange mechanism, 191
  transport, 146, 155, 169, 180, 191–193, 196, 208
    chain (ETC), 146, 155, 169
Electroneutrality, 181
Endergonic reaction, 148
Endothermic
  process, 40, 57, 119
  reactions, 7, 14, 16, 61, 93
  route, 7
Energy crop systems, 128
Enteric bacteria, 151
Environmental
  pollution, 128, 189
  remediation, 194, 208
Equilibrium, 12, 62, 145
Equimolar ratio, 86
Equivalence ratio (ER), 33, 40, 42, 46, 47, 64, 118, 119
Escherichia coli (*E. coli*), 134, 151, 152, 163–169
Esters, 39
Ethane, 79, 80, 101, 114
Ethanol, 79, 80, 101
Eubacteria, 153
Eubacterial systems, 153
Eukaryotic microorganisms, 21
Exoelectrogenic bacteria, 190, 192, 193
Exoelectrogens, 192
Exothermic
  processes, 57
  reactions, 6
  transformation, 78
Extracellular electron transfer (EET), 192, 195, 209

F

Feedstocks, 7, 10, 13, 34, 35, 39, 55, 78–80, 91, 94, 101, 102, 110, 111, 129, 133, 135, 175
Fermentation, 127
Ferredoxin (Fd), 130, 134, 147, 149–153, 157, 161, 169, 203
Fertilizers, 2
Field-effect transistors (FETs), 193, 209
Fischer-Tropsch synthesis, 87, 177
Fluidized bed reactor (FBR), 15, 16, 41, 42, 44–54, 64, 79, 87, 119
Food
  industry, 2, 78, 111
  waste, 23, 176
Formate, 151, 152, 161, 163–167
  hydrogen lyase (FHL), 151–153, 163–165
Fossil fuel, 2, 3, 25, 26, 33, 34, 63, 77, 110, 128, 129, 135, 141–143, 145, 168, 176, 177, 190, 199, 208
Fourier-transform infrared (FTIR), 20
Fuel cell, 2, 3, 34, 79, 110, 128, 130, 144, 159, 160, 176, 199
Functional groups, 39, 181

G

Gas
  chromatography, 20
    mass spectrometry analysis, 17
  gas reactions, 11
  outlet, 112
  solid reactions, 11
Gasification, 3, 5–14, 25, 26, 33, 35–37, 39–54, 60, 63, 64, 78, 94, 98, 101, 109, 110, 112–123, 129, 143, 145, 146, 177, 199
  agents, 52, 109
  reaction mechanism, 113
Gasifiers, 8, 41, 43, 44
Gasoline, 2, 34
Gene manipulation, 143
Genetic engineering, 141, 153, 154, 160, 161, 169
Genomics, 25, 153
Geobacter, 192, 196
Gibson assembly, 153
Global
  depletion, 142
  primary energy, 34
  temperature, 2
  warming, 77, 101, 176
Glucose, 165, 179
Glycerol fermentation, 166, 167
Glycolysis, 133, 152, 165
Golden gateway cloning, 153

Green
algae, 21, 22, 130
energy, 110, 144
Greenhouse, 2, 55, 63, 77, 86, 92, 110, 142, 176, 177, 190, 208
effect, 2
gases, 2, 77, 86, 92, 142, 176

H

Habitation, 110
Heat balance, 57, 101
Heating value (HV), 8, 11, 40–42, 46, 47, 110, 113, 119, 120
Hemicellulose, 17, 42, 43, 111, 114, 133, 162
Heterocyst, 148, 149
differentiation, 149
Hierarchical porous carbon (CPC), 198
High-density polyethylene (HDPE), 111, 116, 117, 123
High-temperature catalytic biomass gasification, 9
Humic acids, 193
Hup-hydrogenase, 148
Hybrid fermentation technology, 23
Hydraulic
retention time, 134, 135
turbines, 142
Hydrocarbon, 2, 6, 11, 13, 26, 34, 36, 39, 44, 56, 62, 78, 91, 98, 109–111, 114–116, 119–122, 128, 143–146
Hydrogen
biological production, 130
metabolism, 21
production, 142, 143
biological processes, 146
electrochemical processes, 143
microbial routes, 177
thermochemical processes, 145
rich gas, 17
Hydrogenase, 141, 142, 154, 156–159, 164, 165, 167–169
enzyme, 21, 22, 133, 141, 147, 148, 180
Hydrogenation, 2, 98
Hydrogenotrophic methanogens, 182
Hydrolysis, 15, 133, 168
Hydroprocessing, 177
Hydrothermal
gasification, 9
method, 50
Hydrotreating, 78, 177

I

In situ CO_2 elimination, 58
Inorganic substances, 132
Intermembrane, 147

K

Ketones, 39

L

Lattice oxygen, 52, 53
Le-Chatelier's principle, 79
Lifecycle assessment, 135
Ligation free cloning, 153
Light fermentation, 130
Lignin, 9–11, 17, 39, 42, 43, 111, 114
Lignocellulose, 42
Lignocellulosic
biomass, 36, 128, 176
materials, 133
matrix, 10
Liquefaction, 5, 36
Liquid hourly space velocity (LHSV), 16, 26
Low
density polyethylene (LDPE), 111, 123
lower heating value (LHV), 8, 41, 49, 50
temperature catalytic biomass gasification, 9

M

Mass
fraction, 59
spectrometry techniques, 20
Maximum molar fraction, 51
Menaquinone, 150
Mesoporous silica carbon (MSC), 87
Metabolic
engineering, 142, 143, 149, 162, 164, 165, 175
flux, 154, 158, 161, 162, 164–167
pathways, 164, 203
Methanation reaction, 7, 9, 51, 114
Methane, 23, 36, 46, 47, 55, 61, 62, 78–80, 87, 88, 90–95, 97, 101, 135, 145, 182, 199, 204
autothermal reforming, 91
combined reforming, 92
dry reforming, 86
partial oxidation, 87
tri-reforming, 93

Methanogenesis, 23, 204
Methanogenic bacteria, 134
Methanol, 9, 13, 16, 55, 78–80, 87, 91, 98, 101, 110, 199
Methylotrophs, 134
Microbes, 3, 23
Microbial
  cell, 130, 141–143, 153, 163, 168
  consortia, 180, 181
  dialysis cells (MDC), 178, 185
  electro synthetic cells, 190
  electrocatalyst, 192
  electrochemical system (MES), 178, 185, 189–191, 194–197, 199, 208, 209
  electrolysis cell (MEC), 130, 131, 136, 175–185, 190
  electrolytic cells, 130, 190, 194
  fuel cells (MFC), 160, 178, 180, 185, 190, 195, 197, 198, 209
  metabolism, 177
  strains, 25
  system, 142, 143, 153
Micro-gas chromatography techniques, 17
Microwave
  plasma, 13
  pyrolysis, 20
Molar
  ratio, 12, 57, 62, 91, 92
  stoichiometry, 166
Molecular
  biology, 142, 161
  energy, 150
  engineering, 153, 168
  events, 21
  hydrogen, 2
  machinery, 141, 142, 153
  reaction, 148
  system, 143
  tools, 153, 154
  weight, 22, 91, 114
Monomeric units, 22
Monomers, 22
Multi-dimensional sectors, 127
Multifunctional catalyst, 60
Multi-redox cycles, 58
Municipal solid waste (MSW), 3, 23, 34, 109, 117, 128, 176–178

N

Nafion, 181, 182
Nanocatalyst, 10, 191, 197, 199, 200, 202–204
Nanocomposite, 195, 196
Nanoelectrode, 194–198, 209
Nanomaterials, 194
Nanomodification, 191, 194
Nanoparticles (NPs), 183, 189, 194, 195, 197, 200–205
Nanoscience, 208
Nanowires, 192, 193, 195, 196
Naphtha, 79
Naphthalene, 40
National renewable energy laboratory (NREL), 16
Native microbial cell systems, 143
Next-generation sequencing, 153
Nickel
  iron layered double hydroxide (NiFe LDH), 197
  lanthanum oxide catalysts, 80
  oxide (NiO), 18, 19, 49, 50, 53, 60, 62, 91, 121, 198, 203
  platinum catalyst, 80
Nicotinamide adenine dinucleotide (NAD), 134, 136, 148, 150, 153, 163, 203
Nitrogen fixation, 141, 148, 149, 153, 157, 160
Nitrogenase, 142, 148, 150, 153, 154, 157, 158
Noble metal, 79, 87, 91, 101
  catalyst, 17
Non-biodegradable petrochemical derivatives, 176
Non-catalyst drenched biomass, 10
Non-conventional
  electrodes, 198
  fuels, 110
Non-isothermal experiments, 60
Non-oxygenic pathway, 149
Non-renewable sources, 2
Non-sulfur bacteria, 132, 150, 151
Non-thermal plasma, 13
Non-uniform biomass properties, 41
Non-volatile substances, 6
Novel sorbents, 63
Nuclear
  fusion
    energy, 14
    reactor, 14
  power systems, 14

O

Ohmic losses, 191
Olivine, 47, 48
Optimal microbial cultures, 190
Organic
  acids, 22, 132, 133, 135, 151, 165
  carbon, 51
  matter, 130, 145, 146, 178, 180, 183, 190
  molecules, 22, 150, 151, 162
  municipal waste, 111
  sludge, 7
  substrates, 22, 23, 178, 203
  wastes, 34, 112, 128, 132, 178
Oxidation, 17, 19, 21, 57, 58, 61, 87, 91, 93, 96, 113, 114, 119, 135, 149, 151, 153, 179–181, 190, 196, 198, 199
  zone, 113, 114
Oxidative cracking, 16
Oxygen, 3, 5, 8, 12, 14, 19–22, 33, 34, 36, 39, 41, 52–54, 57, 58, 61–63, 91, 94, 97, 98, 111–114, 119, 130, 132, 133, 135, 143, 145–148, 150, 155, 158–160, 180, 182, 192, 194, 195, 209
  carrier (OC), 19, 33, 52–54, 57, 58, 61, 62
  reduction reaction (ORR), 194, 199, 209
  transfer material (OTM), 62

P

Partial oxidation (POX), 5–7, 11, 17, 40, 54, 55, 57, 78, 87, 90, 91, 93, 98, 101, 102, 112, 114, 145
Petroleum, 2, 10, 77, 78, 110, 143
pH, 23, 131, 134, 135, 166, 178, 181, 182
Phenolic compounds, 39
Phenols, 39
Phosphoenolpyruvate carboxylase, 166, 167
Photo bacteria, 146
Photoautotrophic microorganisms, 130
Photo-catalyst, 101
Photochemical methods, 129
Photo-electrochemical processing, 129
Photofermentation, 21–23, 127, 129, 131–136, 144, 146, 178, 200, 202, 204
Photolysis, 21, 22, 35, 98, 130, 135, 141, 142, 146, 153–155, 200
Photons, 147, 150
Photoproduction, 147
Photoreaction, 146
Photosynthesis, 3, 21, 22, 130, 132, 146, 147, 149–151, 155, 158
Photosynthetic
  bacteria, 22, 23, 132, 135, 150
  pigments, 130
  protons, 21
  reactions, 146, 147
Photosystem II (PS-II), 21, 22, 132, 147, 149, 169
Phycobiliproteins, 130
Phycobilisomes, 147
Phycocyanin (PC), 147, 150, 169
Phycoerythrin (PE), 111, 117, 118, 120–122, 147, 169
Plasma, 13, 41
  aids, 13
  arc decomposition, 13
Plastocyanin, 150
Plastoquinone (PQ), 147, 150, 169
Pollutants, 116, 130, 190
Poly(3,4-ethylene dioxythiophene) (PEDOT), 198
Polyaniline, 198
Polyethylene (PE), 111, 120, 123, 198
  terephthalate (PET), 111, 115–118, 120, 123
Polymer, 6
  electrolyte membrane electrolyzer, 143
Polymerase chain reaction (PCR), 153
Polymeric molecules, 6, 9
Polymerization, 10, 40, 192
  rate, 10
Polyolefines, 116
Polypropylene (PP), 111, 116, 121–123
Polystyrene (PS), 111, 116, 120, 122, 123, 147, 149, 150
Polyurethane, 111, 123
  polyvinyl chloride (PVC), 111, 117, 122
Polyvinylchloride, 123
Primary catalyst, 47, 48
Principal chemical reaction, 14
Proteomics, 25
Proton, 146, 147, 150, 151, 178–181, 202
  exchange membrane (PEM), 178, 180, 181, 185
Pyrolysis, 3, 5, 6, 12, 14–17, 19, 20, 25, 26, 33, 36, 37, 39–41, 44, 47, 50, 52, 54, 63,

78, 94, 97–99, 101, 110, 113, 114, 121, 143, 145–177
parameters, 36
reaction, 98
Pyrolytic
char, 113, 114
products, 114, 121
Pyruvate
ferredoxin oxidoreductase (PFOR), 134, 152, 153, 161, 162
formate lyase (PFL), 152, 161, 163

R

Raw materials, 2, 26, 55, 78, 80, 111, 146
Reactant ratio, 87
Reaction
calcination cycles, 60
center-1 (RC-1), 150, 169
time, 10, 41, 49, 134, 135
Redox
enzymes, 192
reactions, 147, 197
Reduced ferredoxin (Fd_{red}), 130, 153
Reforming, 3, 5, 9, 11–13, 15–20, 25, 26, 33, 34, 36, 44–50, 53–59, 61–64, 77–81, 86–88, 91–98, 101, 102, 114, 115, 121, 129, 142, 143, 145, 146, 177, 199, 202
Renewable
energy, 3, 25, 34, 143, 145, 176, 196
resources, 34, 35, 63
sources, 129, 175
Rice husk
ash, 50
char (RHC), 50, 54
RNA sequencing technology, 153
Ruthenium, 9

S

Saccharification, 133
Secondary
catalyst, 47, 48
hydrocarbons, 13
Semiconductor, 78, 101, 192
Sequential catalytic process, 18
Shewanella, 182, 192, 194, 195, 198
Socio-economic benefits, 2, 177
Sodium carbonate, 9
Solar
chemical energy conversions, 199
energy, 12, 13, 21, 22, 145–147, 150, 176, 184
gasification, 13
reactor systems, 13
Solid oxide electrolyzer, 143
Sorbents, 35, 42, 59, 61, 63
Sorption enhanced (SE), 19, 33, 58, 61–63
chemical looping steam reforming (SE-CLSR), 19, 61–63
steam reforming (SESR), 58–61, 63
Steam
carbon reaction, 7
gasification, 11, 42, 44, 116, 121
reforming (SR), 9, 13, 15–17, 19, 20, 25, 26, 33, 34, 36, 44, 46, 47, 49, 54–59, 61–63, 77–81, 87, 91–93, 97, 98, 101, 102, 115, 129, 199
process, 54, 80, 87, 97, 98, 101
to-biomass (S/B), 6, 42, 44–47, 53, 64, 118
ratio, 64
to-carbon (S/C) ratio, 16, 26, 56, 59, 60, 62
to-fuel (S/F), 80, 118
Stoichiometric weight, 118
Stoichiometry, 6, 162
Subcritical hydrothermal gasification, 10
Substantial energy system, 3
Supercritical
hydrothermal gasification, 11
water, 9, 10, 12, 25, 51
gasification (SCWG), 9, 10, 12, 25, 50–52, 63
Sustainable energy sources, 1, 3, 25
Syngas, 6–8, 10, 13, 20, 40–42, 46, 47, 50, 52–55, 57, 78, 79, 86, 87, 91, 92, 94, 115–120, 122, 145

T

Tar elimination, 50
Thermal
decomposition, 6, 14, 35
energy, 7, 13, 17, 25, 112
gasification, 40
isolation, 19
reforming, 91
Thermo-catalytic conversion, 64
Thermochemical

conversion, 1, 4, 5, 25, 26, 110, 113, 128
route, 4
limitations, 23
methods, 25, 26, 35, 142, 177
processes, 3, 6, 21, 25, 26, 36, 42, 63, 129, 146
reactions, 39
routes, 1, 3, 5, 22, 33, 56
techniques, 109, 110
Thermodynamic
limitations, 33, 63, 200
method, 52
Thermogravimetric
analysis (TG), 54
analyzer, 17, 52
Fourier transform infrared spectroscopy (TG-FTIR), 54
Thermolysis, 98, 101
Thermophysical properties, 9
Thylakoid membrane, 148
Toluene, 40, 49
Transcriptome analysis, 153, 167
Tri-reforming of methane (TMR), 78, 93, 94, 102
Troubleshooting, 25

U

Ultrafiltration membrane, 181
United Nations sustainable development goals, 128

V

Valorization, 7, 40, 42, 50, 55
Vegetative cells, 147, 149
Volatile
acids, 133
fatty acids (VFA), 131, 136
matter content, 112
substances, 11
Volatilization, 16

W

Waste
biomass, 109–111, 127, 135, 136, 143, 162, 166
plastics, 109, 111, 119, 122, 123
Wastewater remediation, 194
Water
desalination, 178
gas shift (WGS), 5, 12, 15, 19, 33, 35, 39–41, 43–45, 47, 51, 56–59, 61, 63, 79, 86, 87, 113–115, 120, 122
reaction, 7, 12, 40, 43, 86, 113, 114, 120, 122
splitting
method, 98
photosystem, 150, 155
reaction, 100, 101
Wetness impregnation method, 51

X

Xerogel, 51

Z

Zero green-house gas emission, 145
Zone combustion, 8

For Product Safety Concerns and Information please contact our EU representative GPSR@taylorandfrancis.com
Taylor & Francis Verlag GmbH, Kaufingerstraße 24, 80331 München, Germany

www.ingramcontent.com/pod-product-compliance
Lightning Source LLC
LaVergne TN
LVHW050521100826
845148LV00002B/406

* 9 7 8 1 7 7 4 6 3 9 8 1 8 *